THE ILLUSTRATED BOOK
OF VERTEBRATES

ACKNOWLEDGEMENTS

This title first appeared as **The Oxford Book of Vertebrates** and Peerage Books gratefully acknowledge the co-operation of the Oxford University Press who gave permission for this edition to be published.

THE ILLUSTRATED BOOK OF VERTEBRATES

THE MAMMALS, FISH, REPTILES AND
AMPHIBIA OF BRITAIN

Illustrations by

DEREK WHITELEY

Text by

MARION NIXON

First published in Great Britain in 1972 by
Oxford University Press as
The Oxford Book of Vertebrates

This edition published in 1985 by
Peerage Books
59 Grosvenor Street
London W1

© Oxford University Press 1972

ISBN 1 85052 023 2

Printed in Czechoslovakia
50581

Contents

	Page
INTRODUCTION	vi
CYCLOSTOMES	viii
Lampreys and hagfish	viii
CARTILAGINOUS FISH OF THE SEA	2
Sharks	2
Rays	10
BONY FISH OF THE SEA	20
Pelagic teleosts	20
Deep water and oceanic teleosts	40
Bottom-living teleosts	44
Off-shore teleosts	56
Inshore teleosts	60
Inshore and estuarine teleosts	74
BONY FISH—MIGRATORY FORMS	22, 84–90, 104, 106
BONY FISH OF FRESH WATER	90
AMPHIBIANS	114
REPTILES	120
Marine turtles	120
Terrestrial forms — lizards and snakes	122
TERRESTRIAL MAMMALS	126
Insectivores	126
Rodents	130
Rabbits and hares	138
Deer	142
Sheep, cattle and goats	150
Ponies	152
Bats	154
Carnivores	164
AQUATIC MAMMALS	176
Carnivores	176
Cetaceans	182
THE VERTEBRATES — A CLASSIFICATION	192
THE VERTEBRATES OF THE BRITISH ISLES	199
GLOSSARY	204
UNITS OF MEASUREMENTS	205
SOURCES OF FURTHER INFORMATION	206
INDEX	207

INTRODUCTION

The vertebrates first appeared 500 million years ago, some 4500 million years after the beginning of the formation of the earth. They now comprise about 51,000 species of an estimated total 1,000,000 known living species of animals.

Until 350 million years ago the vertebrates were all aquatic, but now they display great diversity of form illustrating the many modifications that have occurred during the transition from a totally aquatic life to a terrestrial one. Despite this vast radiation there are many basic structural features common to all although some are only seen in the embryonic stages.

These basic features involve both the skeleton and organs of the animal. The *backbone* is dorsal and supports the animal; it can either be a longitudinal rod, the notochord, or a set of jointed vertebrae made of bone or cartilage or a combination of both. From this rod arise *two girdles*, the pectoral and pelvic, that support the fins, or limbs, as well as the muscles which make them work. The *nervous system* lies close to the notochord and is a dorsal, hollow tube expanded at the front end to form a brain. It is usually protected throughout its length by skeletal elements. The *respiratory system* develops from the pharynx and in the adult shows diverse forms from gills to lungs. The single heart, a multichambered organ, pumps the blood, containing red corpuscles, around the body. This is the major transport system of metabolic substances throughout the tissues. Like the respiratory system the *digestive system* shows many adaptations for special diets. Generally it consists of a mouth, jaws, teeth, oesophagus, stomach(s), small and large intestine, and anus. The liver, an organ common to all vertebrates, is important in the utilization of the products of digestion.

In their attempts to survive in new habitats the vertebrates have undergone many striking alterations in their shape, structure and functional organization. When the fossil vertebrates are considered together with the living ones, they reveal a continuing story of modification and dramatic change linked to the conquest of new environments. The primitive, paddle-like fin has become variously altered to perform efficiently such diverse activities as walking, running, jumping, climbing, grasping, swimming and even flying. Lungs for breathing air have developed instead of the gills used for respiration in water. The skin has become less permeable to prevent water loss as animals have come to spend more time on land. Compared with the early, primitive vertebrate brain, the relatively large, convoluted one found in mammals indicates the considerable increase in its complexity necessary for the success of these animals in the difficult land environment. The body temperature of most vertebrates is close to that of their surroundings but the birds and the mammals are able to maintain theirs, irrespective of their environment, at about 37°C. This has enabled them to invade many new habitats even in arctic and antarctic climates. Many vertebrates, particularly bony fish, produce millions of small eggs but there has been a tendency towards the development, within the mother, of a few relatively large young that are born alive. After birth the offspring usually receive parental care and protection, and therefore have a better chance of survival. Thus besides the many external changes there have been innumerable internal ones.

The following pages survey the wide variety of vertebrates found in and around the British Isles and where possible the illustrations show them in their habitat. The sizes of the animals are given in the text; the record sizes given for fish caught on rod and line are all for those taken in England, Scotland, Northern Ireland and Wales; those captured in Eire are shown as the Irish record fish. The scientific name has been given for all the animals (those in parentheses are considered to be synonyms of the valid name preceding) and the common name for most.

Nuptial pad on fore-foot
of the male common frog

Isolated dermal denticle from
the skin of a dogfish

Isolated cycloid scale from a herring
of 7 years, showing 7 winter rings

1. CYCLOSTOME
Lamprey

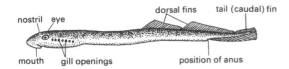

nostril · eye · dorsal fins · tail (caudal) fin · mouth · gill openings · position of anus

2. CARTILAGINOUS FISH
(a) **Spiny dogfish**

1st dorsal spine · 1st dorsal fin · tail (caudal) fin · 2nd dorsal spine · 2nd dorsal fin · spiracle · eye · gill slits · pectoral fin (paired) · pelvic fin (paired) · lateral line

(b) **Ray** (flattened dorso-ventrally)

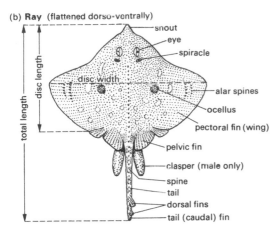

snout · eye · spiracle · disc width · alar spines · ocellus · pectoral fin (wing) · pelvic fin · clasper (male only) · spine · tail · dorsal fins · tail (caudal) fin · total length · disc length

3. BONY FISH
(a) **Herring**

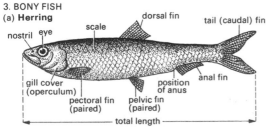

dorsal fin · tail (caudal) fin · scale · nostril · eye · gill cover (operculum) · pectoral fin (paired) · pelvic fin (paired) · position of anus · anal fin · total length

(b) **Plaice** (flattened laterally and lies on one side)

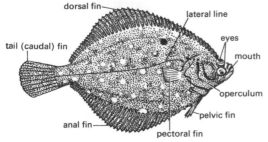

dorsal fin · lateral line · eyes · mouth · tail (caudal) fin · operculum · pelvic fin · anal fin · pectoral fin

4. AMPHIBIAN
Frog

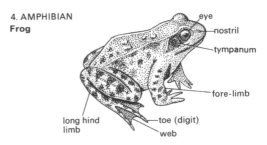

eye · nostril · tympanum · fore-limb · long hind limb · toe (digit) · web

5. REPTILE
Snake

eye · nostril · small scales · large scales · tongue with forked end

6. MAMMALS
(a) **Wild cat** (carnivore)

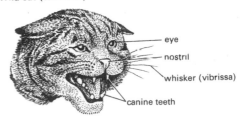

eye · nostril · whisker (vibrissa) · canine teeth

(b) **Horseshoe bat** (adapted for flight)

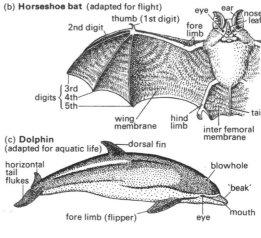

2nd digit · thumb (1st digit) · fore limb · eye · ear · nose leaf · digits { 3rd 4th 5th · wing membrane · hind limb · inter femoral membrane · tail

(c) **Dolphin** (adapted for aquatic life)

dorsal fin · horizontal tail flukes · blowhole · 'beak' · fore limb (flipper) · eye · mouth

(d) **Red deer stag** (ruminant)

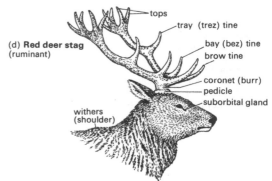

tops · tray (trez) tine · bay (bez) tine · brow tine · coronet (burr) · pedicle · suborbital gland · withers (shoulder)

LAMPREYS AND HAGFISH

These most primitive of living vertebrates are slimy and eel-like, lack paired fins, and are characterized by a notochord, a cartilaginous skeleton, and a mouth without jaws. They are members of the class Agnatha and the order Cyclostomata. The lampreys (1 – 3) are migratory and spawn, in specially prepared nests, in fresh water; the sea and river lampreys (1, 2) ascend from sea to river and the brook lamprey (3) moves upstream. The eggs hatch into ammocoete larvae (1A) which migrate downstream after metamorphosis. Hagfish (4) are entirely marine. Their large eggs (4B) hatch into small hagfish, with no larval phase.

1 **Petromyzon marinus** (Sea Lamprey). It has rarely been caught in the sea, where it lives as an adult at depths of up to 500 m, but it has been recorded in the western Channel and the Isle of Man. Maximum length about 100 cm, weight 2·5 kg. Larval lampreys, or ammocoetes, live in streams and rivers. Non-migratory populations live in Loughs Conn and Corrib and reservoirs on the River Lee in Ireland.

In spring, the adults migrate up river to spawn in streams with strong currents and a bottom of sand or gravel. The male and the female, by removing stones with their suckers and piling them on the downstream side form a depression in the bed of the stream. Spawning begins in May or June when the water is about 15°C. The female extrudes some eggs which after being fertilized drift to the nest-edge and remain amongst the stones. This procedure is repeated until the adults are spent, and die. An average of some 170,000 small eggs are laid, and they hatch in 10 – 12 days.

About 20 days after hatching the larvae drift to quieter waters where they remain in a burrow until metamorphosis. This begins in autumn when the animals are about 5·5 years, 13 to 16 cm long, and is completed by mid-winter. The lamprey then migrates to the sea where it remains until it returns to the river to spawn and die.

2 **Lampetra fluviatilis** (Lampern or River Lamprey). They can attain a length of 40 cm but average 25 – 30 cm. Lamperns used to be captured in large numbers in the Severn for use as food or bait. The adults spend about a year in the sea around the coast feeding on crustaceans, worms, and dead fish, and attacking live fish.

In autumn and spring they migrate into rivers. They spawn in the following April when the male glides along the female using his sucker, attaches himself to her head, and then winds his tail around her. The female shakes her body during the breeding act which may last only 5 seconds. About 100 eggs are extruded and fertilized by the sperm, which are released simultaneously with the eggs and remain active in freshwater for only about 50 seconds. The procedure may be repeated many times. Nest building takes place between spawning. After laying about 14,000 – 26,000 eggs both adults are spent and soon die.

The larvae are now believed to remain in the river for 4·25 years on average, metamorphosis taking place when they have reached a length of 8 – 12 cm. A further

6 – 8 months are spent in the river before migration to the sea in the autumn and winter. During this time the lampern stops feeding.

3 **Lampetra planeri** (Brook Lamprey). They can reach a length of 17 cm and a weight of 11·4 g, but average 13 cm and 6 g. Common in the British Isles they are probably present in most unpolluted streams and rivers except in the north of Scotland. Adults spawn in March or April, and when the nest is completed the male moves his oral disc along the dorsal surface of the female. Both then arch the anterior portion of their bodies as the male coils his tail around the female. Rapid movements are then made back and forth against the substratum till eggs and sperm are extruded, a small number at a time, into a 'cloud' of sand grains to which the eggs adhere. The whole procedure is repeated until 1000 – 2000 ovoid eggs, 1·1 mm long, are laid. After about 10 days, at 10 – 15°C, the eggs hatch. The ammocoete larvae drift to a protected site well stocked with food. They measure about 2·6 cm at 6 months, 7 cm at 2·5 years, and 14 cm at 5·5 years. Metamorphosis depends on size and age, but usually takes place at 6·5 years, occupying July till early winter. The adult does not feed, and dies after spawning. They live for about 7 years.

4 **Myxine glutinosa** (Hagfish, Borer) is 14 to 25 cm long, the female being slightly the longer. It is rare in British waters — one caught off the Isle of Man 1926, one reported in the Moray Firth, 1952. It lives entirely in the sea between 30 and 900 m but usually below 100 m. Underwater observations show that a 30-cm hagfish can swim at 1 m per second over a distance of 10 m. Generally it lies partly buried in soft mud but swims with snake-like undulations to find food, which may be injured or dead fish. This it enters via the mouth or anus where the irregular surface makes penetration easier. It sometimes twists itself into a running knot to exert tension (4) while taking fragments of large prey. Other small soft-bodied animals are also eaten.

In the young, the germinal elements of both sexes are combined in the reproductive organs which only later develop into a testis or an ovary. The mode of reproduction is unknown. Females lay 1 to 22 large, yellowish, elliptical eggs with a horny shell, each about 2·5 cm long. At hatching the young, about 4·5 cm, resembles an adult.

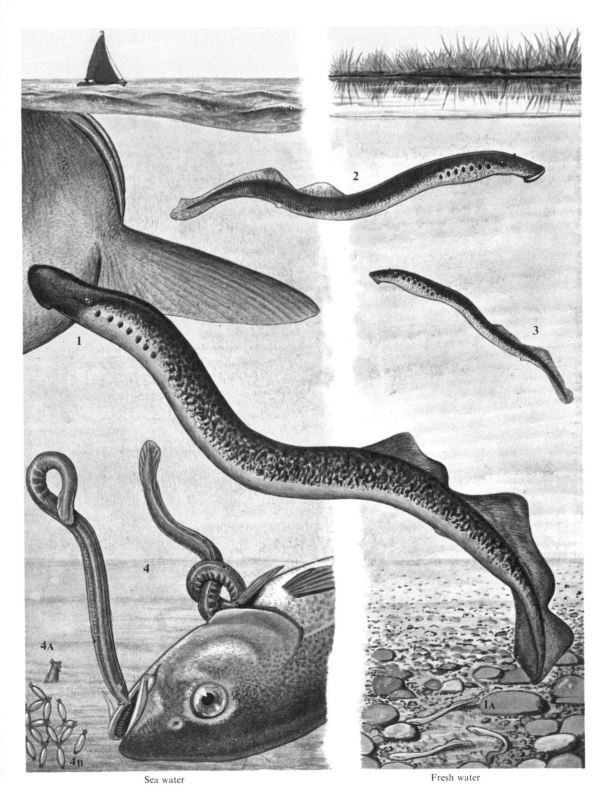

Sea water

Fresh water

1 SEA LAMPREY, attached to basking shark

1A SEA LAMPREY, ammocoete larvae

2 LAMPERN

3 BROOK LAMPREY

4 HAGFISH, entering mouth and gills of dead fish

4A HAGFISH, head protruding from sand

4B HAGFISH, eggs

1

SIX-GILLED, FRILLED AND MACKEREL SHARKS

Fish of the class Chondrichthyes characteristically have jaws, a cartilaginous skeleton, skin covered with denticles, and paired pectoral and pelvic fins, the latter bearing claspers in the male. The majority are marine, as are all the British species. Sharks belong to the order Selachii. They have a rounded body, several gill-openings on each side, and a well-developed tail fin with a long upper lobe. Two sharks on this page (3, 4) are primitive and distinguished by six pairs of gills. Mackerel sharks (1, 2) are large oceanic migratory fish.

1 **Lamna nasus** (Porbeagle) is an active, pelagic, oceanic shark found around the British Isles in summer. It can sometimes be seen at the surface and its range appears to be from the surface to depths of 180 m. The porbeagle and mako have been confused. The porbeagle has a stouter body and the height at the base of the pectoral fin is 14·5 per cent of the total length; in the mako this ratio is 11·3 – 11·9 per cent. It can reach a length of some 2·7 m; the record rod-caught one weighed 195·036 kg. The gape of the mouth is large and the teeth, similar in both jaws, are used for seizing and holding the prey which is swallowed whole. Food is mainly fish — mackerel, herring, cod, dogfish — and also squids.

The female is sexually mature when about 1·5 m long and is ovoviviparous. After emerging from the egg, the embryo remains in the uterus without connection, and develops there. It obtains nourishment by swallowing unfertilized eggs in its vicinity, and this results in a large 'yolk stomach'. Between 2 and 5 young are born: one female captured had 3 embryos 45 cm, 47 cm, and 60 cm long; another, weighing 166 kg, had 3 young, 9 – 11 kg, almost 100 cm in length. Some very small porbeagles captured off Looe in the summer were probably born in British waters.

2 **Isurus oxyrhinchus** (Mako or Sharp-nosed Mackerel Shark). The first recorded in the British Isles in 1955 was a male, 2·6 m long and 160 kg, captured off the Cornish coast. Since then a number have been caught including two in Irish waters. The record rod-caught specimen, from Eddystone Light, weighed 226·786 kg. A strong, active swimmer, the mako is capable of leaping clear of the water and provides excellent game for fishermen. It usually swims near the surface, and the tips of the first dorsal and the caudal fin may be seen above the water.

The mouth is large and the teeth in each jaw are alike. It feeds on mackerel and herring as well as much larger fish; indeed a swordfish of 54 kg was found in the stomach of a mako of 331 kg. Males reach sexual maturity, indicated by the development of the claspers, when about 180 cm long, but females probably not until they are larger. Little is known of their reproductive habits. Like the porbeagle (1), the young have large 'yolk-stomachs' and reach considerable size before birth. A female of 141 kg, captured off Looe, contained a single embryo close to full development.

3 **Hexanchus griseus** (Six-gilled Shark) is a visitor from warmer waters that sometimes occurs in British waters at depths of 180 – 1100 m.

It belongs to the Hexanchoidea, a group of sharks with many primitive features including an unrestricted notochord and six gill-slits. The head is broad and the large eyes, nostrils and gills set on each side are widely separated. The nostrils and large mouth are on the ventral surface. The mouth has a wide gape and the teeth of the upper jaw have a single large cusp with one or two smaller ones; those of the lower jaw are long with a series of small cusps giving the appearance of a comb, and so named 'comb-teeth'. Its food consists of fish such as hake, as well as crabs, shrimps and cephalopods; indeed a whole torpedo was found in one stomach! This shark probably comes to the upper layers at night to feed, as it has been seen at the surface, but remains on the bottom during the day when it is taken in trawls.

It probably reaches sexual maturation when 200 to 250 cm long. It is ovoviviparous and a large number of embryos can develop in one female; a specimen 480 cm long (not captured in British waters) had 47 embryos. At birth the young are 40 – 66 cm long.

Males and females of 62 – 305 cm have been taken and one small one of 61·5 cm weighed 840 g. A large one of 70 kg was captured on rod and line off Kinsale, Co. Cork, recently.

4 **Chlamydoselachus anguineus** (Frilled Shark). This rare animal is a living representative of a primitive group. First described in 1884, it had by 1935 hardly been taken outside Japanese waters. However, several have now been caught in the eastern Atlantic, including four off the British Isles at depths of 560 – 700 m, the largest one 182 cm long.

The colour, never described in a freshly caught specimen, is after preservation dull brown and slightly paler on the ventral surface. There are six pairs of gills. The first gill cover meets to form a continuous fold across the throat, giving the appearance of a frill. The mouth is large and armed with sharp teeth, each with three slender re-curved cusps, arranged in rows extending obliquely from front to back across each jaw. This species is ovoviviparous, the large eggs, 11 – 12 cm long, developing in the female. Between 3 and 12 eggs may be present in one female and the gestation period is long, perhaps two years, for the embryos become very large. The young are born alive and may be 25 – 35 cm at birth.

1 PORBEAGLE   2 MAKO SHARK
3 SIX-GILLED SHARK   4 FRILLED SHARK

LARGE SHARKS

These large pelagic sharks, of the order Selachii (p. 2), may be seen at the surface of the sea although they also go to considerable depths. The basking shark (3) is the largest fish in British waters, second only to the largest fish in the world, the whale shark.

1 **Alopias vulpinus** (Thresher Shark) may reach 5 m in length but those found off the west coasts of Scotland, Ireland and England are usually 4 - 4·5 m. The record rod-caught specimen weighed 127 kg. This shark is characterized by the enormous length of the dorsal lobe of the tail.

The jaws of this predator both have flat, triangular, unserrated teeth of equal size. It drives schooling fish, such as pilchards, mackerel, and herring, into a tight group by swimming round them in ever diminishing circles, and then with its long tail stuns or kills the victims. One thresher's stomach contained 27 mackerel.

Sexual maturity is reached when they are about 427 cm long. Little is known of their reproduction, but they are probably ovoviviparous. One female of 472 cm was found with two young, each 150 cm long. There are usually 2 - 4 pups to a litter. The pup's tail is proportionately even longer than the adult's.

2 **Somniosus microcephalus** (Greenland or Sleeper Shark) occurs fairly often in Scottish waters. This member of the genus Somniosus, which contains the only sharks to inhabit Polar waters throughout the year, generally lives beneath the ice there. Usually about 6 m long and 1100 kg in weight, the largest captured in British waters was 6·4 m and 1020 kg. The Eskimos, at Angmagssalik in Greenland, catch this shark on line and hook through a hole in the ice. It is fished for its liver oil, but the flesh is toxic and is edible only after boiling in several changes of water. Although apparently rather inactive, as the name 'Sleeper' implies, it takes a wide variety of active prey including many fish and squids. Little is known of its reproductive habits but it does bear live young. A litter of 10 pups, 38 cm long, has been reported from one female of 5 m.

3 **Cetorhinus maximus** (Basking Shark) is found off the west coast of Scotland and Ireland from April to October, with occasional reports from November to March. It can be seen in summer swimming slowly near the surface, the large triangular dorsal fin showing above the water. The longest captured around our shores was 12·8 m long; one of 9 m weighed 3500 kg. When captured, sea lampreys (p. 1) are often found attached; they are unable to penetrate the tough skin of the shark, but may leave superficial marks.

This shark's most striking features are the large mouth and enormous gill-clefts, the first pair of which almost meet in the ventral mid-line but are separated by about 15 cm dorsally. Both inner sides of each gill have a line of gill rakers, or combs, which are erected to form an interlocking sieve when the mouth is open. The mouth and gills together form the mechanism by which the minute organisms, including crustacea and fish eggs, are filtered from the plankton. A feeding shark cruises slowly, 3·7 km per hour, with mouth agape; as much as 2215 m³ of water may pass through the gills in one hour. These large sharks feed only for part of the year as the gill rakers are lost in October. New ones develop during winter and are complete by February, this perhaps being the reason these animals are rarely seen in winter which is probably spent in deep water.

The mode of reproduction is still uncertain. Anatomical evidence suggests that it bears live young. Sexual maturity is reached in 3 - 4 years when about 6 m in length.

4 **Prionace glauca** (Blue Shark) has been caught in Scottish waters, and as a visitor to the south coast, from June until late autumn, it provides sport for sea anglers off Cornwall. The tip of the dorsal and tail fins may be seen breaking the surface of the water. A length of 600 cm can be attained but those caught in British waters have been 137 - 229 cm. A stranded one of 190 cm weighed 50·8 kg, while the record rod-caught specimen weighed 98·878 kg. This shark preys mainly on fish such as herring and mackerel, and also eats cephalopods. The blue shark is viviparous. A gravid female, 200 cm, 45·36 kg, captured near Looe, gave birth to 22 young just after capture, three more being found on subsequent examination. The total weight of the 25 pups was 3·06 kg, one male was 35 cm and 75 g, while a female was 38 cm and weighed 89 g. Other adult females of 250 - 300 cm have been found with 28 - 54 pups. Recently born pups have also been found off the south coast. The young are delivered tail first and the 'placenta' is expelled separately. The umbilical cord that joins the embryo to the placenta appears to be absorbed before delivery. The umbilical opening of the young, on the ventral side between the pectoral fins, is not completely closed at birth. The young may be 100 cm long by the end of the first season and be sexually mature when 200 cm. They may reproduce in the third or fourth season. The gestation period, although unknown, is probably of 10 months' duration. Parturition takes place some 80 km off-shore, from April to August in the eastern Atlantic. Some specimens bear scars which are probably acquired during courtship.

There appears to be segregation of the sexes in this species as with some other sharks. Those captured earliest in the year off the south coast are almost all females.

1 THRESHER SHARK   2 GREENLAND SHARK
3 BASKING SHARK   4 BLUE SHARK

SMALL BOTTOM-LIVING SHARKS

These sharks, of the order Selachii (p. 2), live mainly on the bottom in fairly deep water. None reach great size except the bramble shark (5).

1 **Mustelus mustelus** (Smooth Hound, Sweet William) is not uncommon off the south-west coast, living and feeding on the bottom. It is about 123 cm long but may reach 200 cm; the record rod-caught specimen weighed 12·7 kg. It is grey above with pale spots, and white on the ventral surface. The mouth is large but the teeth are flattened, like the rays. It feeds mainly on crustacea but also takes fish such as whiting or mackerel. It is viviparous, the young obtaining nourishment from a yolk sac placenta. A large number of young are usually present and one female was found with 34 embryos, between 19 and 33 cm in length.

2 **Squalus acanthias** (Spur Dogfish) is common around the British Isles. It swims at depths of 10 – 150 m in shoals containing few or many fish similar in size and often of the same sex. As shoals of smaller fish inhabit shallower water, females being larger than males usually inhabit deeper water. In males the claspers, modified parts of the pelvic fins, become enlarged at sexual maturity. Males are then 59 – 69 cm long, 0·8 – 1·0 kg and about 5 years; mature females are 75 –95 cm, 1·5 – 4·0 kg, and 10 years. The spur dogfish is ovoviviparous and carries its young for 2 years. The eggs remain in a horny capsule, usually one in each oviduct; each capsule, or 'candle' contains 2 or 3 embryos. The total number of embryos, varying between 1 and 12, increases with the size and age of the female. A female 73 cm long may bear 3 young, and one of 100 cm, 20 years old, may carry 10. The capsules rupture after 8 to 9 months releasing the embryos into the uterus. The embryos grow at the expense of the yolk sac which is completely utilized by the time of birth when the young are 23 – 31 cm and up to 83 g. The uterus is highly vascular and is important in gaseous exchange and also, probably, in the elimination of waste products. One stock of spur dogfish makes an autumn migration from the Orkney and Shetland Isles to breeding areas off the north-west coast of Scotland. Birth generally occurs near the coast between September and December and the young miniature spur dogfish swim away at once. Two gravid females reported recently, weighing 5·736 kg and 4·017 kg, each had 10 embryos, full term and of total weight 585 g in the larger specimen, but 490 g and still attached to the yolk sacs in the smaller one. Juveniles 30 – 40 cm long have been caught in inshore waters in October when about 8 months of age. Some travel long distances as one tagged off Willapa Bay, Washington, in 1944 was recaptured in 1952 near Japan,

a distance of 7564 km. Tagging has also revealed that it lives for 25 – 30 years.

A strong, sharp spine, present just in front of each dorsal fin, may be used in defence when the animal curls itself into a bow to strike at an enemy. Each spine has a groove containing glandular tissue which secretes a toxic substance. The mouth, on the ventral surface, is quite large. The teeth are of similar shape in both jaws, each having a single, sharp pointed cusp. Food consists of herring, mackerel and pilchard amongst the pelagic fish, as well as demersal ones like whiting, sand eels, dragonets and flatfish. Females, up to 120 cm, are larger than males, up to about 91 cm; ones of 90 – 99 cm weigh about 2·5 kg. The record rod-caught one was 7·739 kg. Of commercial importance as food it is often sold as 'flake'.

3 **Galeus (Pristiurus) melastomus** (Black-mouthed Dogfish) is occasionally found in trawls off the western coasts of the British Isles. In recent years a number have been caught, 30 – 65 cm in length and 143 – 690 g in weight. It is a strikingly marked shark and the inside of the mouth is black. Reproduction occurs throughout the year and the encapsulated egg takes 6 – 9 months to hatch.

4 **Oxynotus centrina** (Humantin) is a rare straggler to our shores from the coast of Portugal. One was taken in 1877. As a group the Oxynotidae seem nowhere abundant.

5 **Echinorhinus brucus** (Bramble Shark), generally described as a ground shark, is an occasional visitor to British waters found at depths of 18 – 185 m. The largest, recorded in 1869, was a female, 2·7 m long, containing 17 eggs; in the stomach were several dogfish some almost a metre long.

Oxynotus paradoxus. The dorsal fin has a single protruding spine. It is a rare species sometimes caught along the Atlantic slope at depths of 340 – 390 m. Of the twelve specimens captured in recent years nine have been females 47 – 89 cm long; one of 60 cm weighed 1·260 kg. Little is known of its reproductive habits but examination of a spent female, 82 cm long, suggests that this species may be viviparous. As they are usually only captured in late spring there may be a migration inshore in order to give birth to the young.

1 SMOOTH HOUND   2 SPUR DOGFISH
3 BLACK-MOUTHED DOGFISH   4 HUMANTIN
5 BRAMBLE SHARK

BOTTOM-LIVING SHARKS

These sharks, of the order Selachii (p. 2), spend most of their life on the sea bed, some in deep water, others in shallow; the hammerhead (2) may sometimes be seen at the surface.

1 **Galeorhinus galeus** (Tope, Sweet William) is fairly widely distributed around the British Isles. It is found in water up to 285 m in depth in winter and early spring, but from April to December at 4 – 37 m where the bottom is of sand or gravel. It feeds on gadoids, flatfish, dragonets, sea bream and wrasse. This species may be ovoviviparous, or viviparous, and as many as 32 young have been found in one female. The gravid females migrate inshore where the young are born and remain during their first winter.
It usually reaches 150 cm and about 32 kg but the record rod-caught specimen was 33·876 kg.

2 **Sphyrna zygaena** (Hammerhead), although rarely captured, is unmistakable because of the wide, hammer-like head. It is sometimes seen with the tips of the large dorsal and caudal fins above the surface of the water. It spends much time on the bottom feeding on skates, sting rays, and sharks, but it is a strong swimmer and also eats herring and mackerel as well as other pelagic fish. Sexual maturity may be reached when about 210 cm long. It is probably ovoviviparous and a gravid female with 37 embryos has been found. The young are 50 cm long at birth. A specimen captured at Ilfracombe in 1865 was 417 cm; another of 381 cm weighed 409 kg.

3 **Dalatias licha** (Darkie Charlie) is a shark of the Atlantic slope, caught in water 178 – 640 m deep. Females, larger than males, can reach a length of 150 cm. It is ovoviviparous, and gravid females have been found with 10 – 16 young. At birth the young are about 30 cm.

4 **Scyliorhinus stellaris** (Bull Huss, Nursehound, Greater Spotted Dogfish) is found all around the British Isles but is common in Irish waters. It lives in deeper water over rougher ground than the lesser spotted dogfish (5). It is oviparous and the egg cases are sometimes found attached to hydroids or seaweeds, such as the brown algae *Cystosira* and also to *Laminaria*, in pools at low spring tides. The egg case is 11 – 12·5 cm long and 4 – 4·5 cm wide. Hatching takes place when the developing young reach a length of 16 cm and while a small remnant of the yolk sac still remains attached. Two young fish, caught in July, weighed 21 g and 46 g. The bull huss feeds mainly on fish, including small specimens of the lesser spotted dogfish and other bottom-living fish, but also on crustacea and cephalopods. It can reach about 150 cm length and 9·5 kg; the record rod-caught specimen weighed 9·610 kg.

5 **Scyliorhinus caniculus** (Lesser Spotted or Common Dogfish, Rough Hound. Two other species are also accorded the name common dogfish, *S. stellaris* and *Squalus acanthias*). It is abundant around our coasts. It does not exceed 70 cm, both sexes being the same size. The record rod-caught specimen weighed 2·041 kg.
A bottom-living species, it has a swimming speed of 24·2 cm per second. Both jaws of the large ventral mouth carry several rows of teeth and as the outer ones become worn they are replaced by new ones that lie behind. They feed mainly on crustacea but molluscs, echinoderms, other invertebrates as well as sand eels, and flatfish are eaten. They reach sexual maturity when about 60 cm long. The claspers, intromittent organs of the male, which are scroll-like modifications of the pelvic fins, do not lengthen markedly at this time. It is oviparous, the eggs being shed after internal fertilization. During copulation (about 20 minutes) the male curls himself around the female and transfers the sperm by inserting one clasper at a time into her cloaca. The eggs are laid in flat, oblong, brown cases, with their corners extended into long threads which are entwined around sea weed to provide anchorage. These are known as Mermaids' purses. Two eggs are usually deposited at one time but many are laid in a season. The breeding season is protracted and although most eggs are deposited in spring, it can occur in any month. In some places along the coast the purses, 5·3 – 6·4 cm in length and 2·1 – 2·9 cm in width, may be collected at the low spring tide. Incubation lasts 157 – 178 days; the embryo develops in the egg case, until it is 9 – 10 cm long, and then emerges with a remnant of yolk sac still attached. In an aquarium the young seem helpless and weak, with the eyes closed and moving only if provoked.

Mustelus asterias (White-spotted Smooth Hound) is found in Irish waters and the western part of the Channel. It is more common than *M. mustelus* (p. 6) with which it is confused, *M. asterias* being distinguished by small white spots on the back and sides. The chief difference lies in the mode of development, for this species is ovoviviparous, the embryo having a distinct yolk sac which has no maternal connection.

1 TOPE
2 HAMMERHEAD
3 DARKIE CHARLIE
4 BULL HUSS
5 LESSER-SPOTTED DOGFISH. male curled round female in copulation
5A LESSER-SPOTTED DOGFISH.egg case

MONKFISH AND ELECTRIC RAYS

The monkfish (1), although a member of the order Selachii (p. 2), has a flattened body and is intermediate between the sharks and the rays. The rays belong to a separate order, the Batoidea, and are characterized by a dorso-ventrally flattened body, very large pectoral fins fused to the head, gill openings on the usually pale underside and spiracles on the dorsal surface. They are mainly bottom-living fish as their shape suggests. The electric rays (2, 3) resemble a flat, rounded disc with a distinct tail and have well-developed sense organs.

1 Squatina squatina (Monkfish, Angel Fish) has been recorded from several parts of the coast. It dwells on the bottom, usually concealed in sand or mud. In summer it comes close inshore but winter is spent in deep water. In form the monkfish is intermediate between the sharks and the rays. The skin is rough with many dermal denticles, the largest being along the mid-line of the back. The underside is white. The mouth is almost terminal and wide, and the teeth sharp, cone-like, and well separated. It feeds mainly on fish, particularly flatfish and mullet, but also eats crustacea and molluscs. Gravid females move inshore to give birth to their young, which are born alive. One female, 105 cm in length, gave birth to 20 young, each 23 – 25·5 cm in length. A length of about 100 cm and a weight of 77 kg can be attained. The British record rod-caught specimen weighed 29·935 kg, and the Irish one 31·298 kg.

2 Torpedo nobiliana (Electric Ray, Crampfish or Torpedo) is found around the British Isles throughout the year, even in northern waters. Of the electric rays this is one of the largest and the one which occurs most frequently in British waters.

The electric organs in the pectoral fins are large and consist mainly of modified muscle tissue. Shocks of 60 volts have been recorded and this organ is used for defence as well as for stunning prey. The movement of a cod or other fish across the visual field of *Torpedo* will provoke it to spring forwards and upwards in attack. Enveloped by the wings and snout the prey, at the moment of contact, is stunned by the shock and then swallowed. As the jaws are distensible, quite large prey can be taken and indeed a cod 50 cm long was taken by a *Torpedo* of 80 cm, but nearly an hour elapsed before the prey was swallowed. The jaws are armed with rows of sharp recurved teeth which make escape impossible. As many as 7 rows of teeth are in use at the same time and there are several replacement rows behind.

Swimming by lateral movements of the tail, it can reach a forward speed of up to 22·5 cm per second. As this fish is only a little more dense than sea water, it does not have to swim continuously. This near neutral buoyancy is due to the large liver (up to 20 per cent of the body weight) and its low density oil (once used in oil lamps). The near neutral buoyancy may be important either when the fish is moving over the soft muddy bottom where it lives, or related to the slow swimming speed, its capacity to turn through 180° with a single stroke of the tail, or its large size. Underwater observations reveal that it is not restricted to living on the sea bed and can be fairly active.

The reproductive habits are largely unknown but it is probably ovoviviparous. The young at birth are 20 – 25 cm long. Two small males 23 cm long and 12 cm across the disc were captured off the Scottish coast at a depth of 27 m in March and June. In each the umbilical aperture was still visible, so birth had occurred only recently. A spent female was also found in the same area indicating that this species must breed off the coast of Scotland. Another recently pregnant female of 15·9 kg was trawled up near Plymouth in May and the oviducts found to be full of uterine milk.

Adults can reach a length of 180 cm and a weight of 50 kg; one specimen captured recently was 100 cm long, 63 cm across the disc, and weighed 9·1 kg. The record rod-caught one weighed 23·586 kg.

3 Torpedo marmorata (Marbled Electric Ray) was first recorded in British waters in 1963. A female, 41 cm long, was speared at a depth of 21 m off the coast of Cornwall. A few other records exist but it seems to be a rare visitor. A length of 60 cm can be attained; one specimen captured recently weighed 3·138 kg. The ventral surface is cream in colour.

This ray is ovoviviparous, the gestation period lasting for 6 months. The young, after release into the uterine cavity, are nourished by uterine milk, a secretion produced by villi which develop on the wall of the uterus. The uterine fluid passes into the digestive tract through the mouth and spiracles, and is digested by the stomach where the digestive glands are functional at an early stage of development. The egg initially weighs about 14 g and the young at birth 27 g.

A shock of about 200 volts can be generated by the electric organ of this ray. During the day the ray remains almost entirely buried in sand but at night it becomes active and swims in search of food. Little is known of its diet except that it generally preys upon fish. The stomach of one specimen contained a three-bearded rockling, *Gaidropsarus vulgaris*, 34 cm long.

1 MONK FISH

2 ELECTRIC RAY   3 MARBLED ELECTRIC RAY

RAYS

The fish on this page belong to the order Batoidea (p. 10) and the family Rajidae which includes the skates and rays. The larger species are generally called skates, and the smaller ones rays. They are characterized by the heart- or rhomboid-shaped disc formed by the two 'wings' and the body. The tail is slender and whip-like. Most are bottom-living and they produce large eggs enclosed in a capsule which is often called a 'mermaid's purse'. All on this page have short snouts and white undersides.

1 Raja clavata (Thornback Ray, Roker) is commonly found around the British Isles, usually to a depth of 185 m but quite often down to 330 m. It prefers rough ground but will inhabit all types of bottom. The eyes, which protrude above the dorsal surface, have some degree of mobility. Large thorn-like spines on the tail and central portions of the wings give this ray its name. These may be worn down in large fish. The majority weigh between 3·2 and 6·8 kg, but the record rod-caught specimen was 17·235 kg.

A shoal often consists of thornbacks of the same sex and size. The very young fish, which hatch in shallow water, remain there in the early stages feeding on crustacea such as amphipods and crangonids. Small and medium size thornbacks live in water of 55 m or less, but after reaching a width of 40 cm across the disc they move to deeper water of 74 – 111 m. Feeding migrations of adults can sometimes be seen in January when fully grown gravid females, almost ready to deposit their eggs, congregate in water of less than 40 m where herring are plentiful. Later the males arrive in ever increasing numbers while the females disappear, finally leaving only males.

Males reach sexual maturity when the disc width is about 50 cm, at 7 years, and females when the disc width is 65 – 70 cm at about 9 years. Males with disc widths of 60 cm are rarely found but females of 80 cm are not uncommon around our shores. Since young fish are present in shallow water at nearly all times, spawning probably occurs almost throughout the year, but egg capsules are encountered most frequently April to July, with a peak in June. Fertilization takes place before the egg is enclosed in its capsule. Each capsule, about 7·4 cm by 5·7 cm, consists of several layers of closely packed keratin-like fibres, has a long horn at each corner, and is covered with a felt-like mass of fibres. The incubation period lasts between 121 and 154 days in an aquarium. When hatched the young are about 13·3 cm long and 8·2 cm wide, and at 4 months are 14·6 cm long and 9·5 cm wide.

The teeth show sexual dimorphism; males have pointed ones and females have flat, rounded teeth, but no differences in their diets have been detected. The adults usually feed on crustacea such as *Corystes* or *Portunus* but sometimes entirely on fish, such as herring, sprats and flatfish. While swimming these fish give an impression of flight, undulating the enlarged pectoral fins, aptly called 'wings', and using the long slender tail as a rudder.

2 Raja montagui (Spotted Ray, Homelyn) is commonly found along the western and southern shores of the British Isles in water of shallow and moderate depths. Generally the adult is about 76 cm long. The record rod-caught specimen weighed 7·342 kg.

The egg capsules, 7·1 cm long and 4·2 cm wide, are found from April to July with a peak in May. The capsule is similar to the thornback's but more delicate and with a smaller mass of attachment threads. After about 5 months' incubation the young at emergence is 12·7 cm long, with a disc width of 7·5 cm, and with irregular black spots on the upper surface.

3 Raja brachyura (Blonde Ray) is frequently found along the Atlantic coast and the Channel but is scarce in the North Sea. It usually lives in depths of less than 120 m and prefers sandy areas. Shoals of this ray often consist of animals which are similar in size and of the same sex. It is not known whether these shoals undertake feeding or breeding migrations. Spawning takes place from March to July. The egg case, about 12·8 cm by 7·8 cm, has a felty mass of fibres on one side. Aquarium observations of a gravid female, kept isolated after capture, showed that eggs were extruded in groups of two, the first followed quite rapidly by the second. There was a period of rest after the extrusion of each pair. The female deposited the eggs April 12 to May 31, but had not then completed her spawning. As each egg was extruded, it was buried in the sand where its fibres became attached to debris. Embryos were present in the capsules when examined, indicating that copulation occurs prior to egg laying, presumably before the formation of the egg case, and there must be a storage site for the sperms within the female. Incubation lasted 189 – 219 days. At hatching the young were 187 mm long and the disc was 115 mm wide. The upper surface was a rich fawn with numerous black spots extending to the margin of the disc.

The young feed largely upon small crustacea such as amphipods, schizopods and crangonids although fish is also taken. The adult takes crustacea and fish such as the herring, razor fish, and sand eels. It can reach a length of 107 cm and a disc width of 76 cm, and may weigh more than 16 kg. The British record rod-caught specimen weighed 16·13 kg, and the Irish one 16·555 kg.

1 THORNBACK RAY   2 SPOTTED RAY

3 BLONDE RAY

RAYS AND SKATES

The fish on this page all belong to the order Batoidea (p. 10) and the family Rajidae (p. 12). All have long snouts, with the exception of the sandy ray (3) which is short-snouted. Of the long-snouted ones, the colour of the underside of the common skate (1) and the long-nosed skate (2) is dark while that of the bottle-nosed ray (4) is white.

1 Raja batis (Common, Blue or Grey Skate) is common in Irish and Scottish waters and around the Isle of Man, but less common in the English Channel and off the south-east coast of England. Until the 1930s it was common in the Channel. Generally to be found at depths of 37 m, it can go down to 370 m. It lives where the bottom is mixed or rough, or of sand or gravel.

The upper surface of this skate is smooth, spines being present on the tail, and of a uniform brown colour except for the black dots that mark the mucus pores. The underside is mottled grey and sepia, the mucus pores again being pigmented. The common skate and the long-nosed skate are the only two 'black-bellied species' found around the British Isles. It may be 203 cm in length with a disc width of 146 cm, and weigh up to 188 kg. The record rod-caught specimen weighed 102·733 kg.

There is a difference between the teeth of males and females. In the male the base is small, the teeth well-spaced and acutely pointed. In the female the base is large, the teeth blunt and set close together. As many as 44 to 55 rows may be present in the upper jaw. Common rays feed upon other species of rays, dogfish, dabs and herrings, octopods and crabs also being taken.

In mating the male and female hold their discs flat. The egg capsules (1A), found in April and May, are about 16 cm long and 8 cm wide. The tips of the horns end as delicate filaments while the whole of the case is covered with a felty mass of fibres. Newly hatched young are about 21·5 cm long and have a disc width of 14 cm. The teeth are closely set, 40 – 45 rows being present in the upper jaw. This skate is commercially important being caught both in trawls and on long-lines.

2 Raja oxyrinchus (Long-nosed Skate) is found off the Atlantic coasts, the south and south-west of Ireland, and the western entrance of the Channel, at depths of 130 – 304 m, occasionally as deep as 925 m. This skate can reach a length of 142 cm with a disc width of 99·4 cm. It is 'black-bellied' (*see* 1), being brown or bluish grey with dark spots and streaks on the underside. It is easily recognized by the elongation of the snout. Both the upper and lower surfaces are covered with spines. The central teeth are sharp and pointed but the lateral ones are round and almost flat; 38 – 42 rows of teeth are present in the upper jaw. Its food consists of crustacea

such as *Cancer*, and fish including *Trigla* and *Callionymus*. The egg capsules are about 13·2 cm long and 8·3 cm wide, with a felty covering of yellowish fibres. Spawning takes place from February to April in the Mediterranean but in the northern North Sea ripe females have been captured in September. The juveniles generally inhabit shallower water, 46 – 101 m, than the adults.

3 Raja circularis (Sandy Ray) is found periodically in the deeper water of the western part of the Channel. It occasionally appears in Irish waters and one caught there recently weighed 1·300 kg and was 59 cm long. The underside is white. Although very little is known of this ray, its egg capsules have been found in April and May; these are about 9 cm long and 5 cm wide. The record weight for a rod-caught specimen is 2·565 kg. A length of 100 cm may be attained.

4 Raja alba (marginata) (Bottle-nosed Ray, White Skate), one of our larger rays, is found in the western part of the English Channel and in Irish waters. Adults generally live in deep water and the young in shallower regions.

The underside is white. Adults may have a tinge of grey on the posterior margin of the wings but in the young there is a broad black border in this region, as well as at the angles of the underside of the wings. The young tend to be reddish-brown on the dorsal surface rather than grey. The mouth has between 40 and 46 rows of teeth set in vertical rows, the outer series being flat and without cusps while the inner series have a long, pointed, posterior, conical cusp. Its food consists of small fish, and crustacea such as *Portunus* and *Galathea*.

The egg capsule is very large, about 18 cm long in the midline and 4 cm wide, one side being more convex than the other. They are generally found in April. One hatched in July after being kept in an aquarium for 15 months. The young fish measured 29·2 cm in length and the disc was 19 cm wide.

The British record rod-caught specimen weighed 34·471 kg, and the Irish record is one of 74·84 kg. Four adults taken recently at Westport, Ireland were 201 – 207 cm in length and 121 – 128 kg. These fish may attain a length of 250 cm and a weight of some 200 kg.

1 COMMON SKATE   1A COMMON SKATE 'mermaid's purse', egg capsule
                   2 LONG-NOSED SKATE

3 SANDY RAY   4 BOTTLE-NOSED RAY

RAYS

The rays on this page belong to the order Batoidea (p. 10), and to the family Rajidae (p. 12). They have white undersides and are short-snouted with the exception of the shagreen ray (4) which is long-snouted.

1 **Raja undulata** (Undulate or Painted Ray) has been found off Devon, Sussex, and Irish coasts in littoral and moderate depths. It has a very restricted distribution. Of specimens caught recently, the two longest were both 91 cm and weighed 7·2 kg and 5·6 kg. The record rod-caught specimen weighed 8·811 kg.

The colouration of the upper surface is quite striking; it is brown, orange or yellow with black or brown undulating stripes, margined by series of small white spots. There are also large white spots scattered on the disc which is covered with spines. The underside is white. The teeth, nearly alike in both sexes, are close set and there are 36 – 46 rows in the upper jaw. It feeds upon fish including plaice, clupeids and gobies as well as crustacea. The egg capsules, found in July off Plymouth, are 8·1 cm long and 5·2 cm wide. The capsule is convex on both sides and is covered by a mass of filamentous attachments on one side, the other side being smooth. Four juveniles have been captured off the Irish coast; the smallest 23·7 cm long weighed 85 g, and the largest weighed about 600 g.

2 **Raja microocellata** (Small-eyed Ray, Painted Ray). It has a very restricted range within its general distribution and can be found in sandy bays and estuaries of the Devon and Cornish coast and the west coast of Ireland from Kerry to Donegal. It inhabits water of shallow to moderate depths. The upper surface and the underside are spiny and the latter is white in colour. A notable feature is the smallness of the eyes. An adult 88 cm long weighs about 5·7 kg. The record weight for a rod-caught specimen is 6·222 kg.

Egg capsules, 9·1 cm by 5·7 cm have been found from April to August. They are smooth and without filamentous attachments, one side being more convex than the other. The long horns are elongated into thin tubes while the others are short and strongly hooked. Incubation probably takes about 7 months, and 4 months after hatching the little ray will be about 19 cm long with a disc width of 9 cm.

3 **Raja naevus** (Cuckoo Ray) is common along the coasts of Devon and Cornwall, and is also present off Ireland, the Isle of Man and Scotland, and in the North Sea, at depths of 150 – 220 m. It is one of the smaller rays found around our shores, reaching a length of 68·4 cm with a disc width of 40·2 cm. The British record rod-

caught specimen weighed 2·268 kg and the Irish one was 2·43 kg.

It feeds on rag worms, sand eels and small crustacea, and probably finds its prey by smell, as it will take bait by night or day. The teeth are pointed in both sexes and 54 – 60 rows are present in the upper jaw of adults.

Ripe females have been found in most months of the year but the majority in the spring. The egg capsule is relatively small, 6·3 cm long and 3·9 cm wide. Convex on both sides and devoid of loose threads, it is more or less transparent yellow changing to reddish brown as development proceeds. Two horns are short and hook-like while the two long ones nearly always cross one another. In an aquarium, after incubating for 243 days, an embryo hatched in February, 11·9 cm in total length, with a disc width of 6·2 cm. Pigmentation was already developed, the prominent mottled ocellus was present, and the tips of the spinulae were beginning to show. After two months the young ray was 12·5 cm long and the disc width had increased to 6·7 cm. About 44 rows of teeth were present in the upper jaw.

4 **Raja fullonica** (Shagreen Ray) is found in the western Channel, the Irish Sea, and the North Sea. Although infrequently taken, this ray has been captured between 160 and 300 m. It can reach a length of 94 cm and have a disc width of 63 cm. The upper surface is covered with spines, while the underside is white. The upper jaw has 68 rows of teeth in the adult and 58 in a juvenile.

Raja richardsoni was first taken in British waters in 1964 from the edge of the continental slope west of the English Channel. It was caught by line at depths of between 2220 and 2405 m. An extremely rare fish, it was first described in 1961, from a single specimen caught off New Zealand. Two fish, a male and a female, were captured in the Channel. The male was 113·3 cm long with a disc width of 78·4 cm, the female was 145·5 cm long with a disc width of 100·4 cm. The male was light brown above and below; the female had clearly defined areas of white around the mouth, nostrils and cloaca. The surface lacked spines but dermal denticles were spread evenly over the whole dorsal surface and over the abdominal area below. The head was rather broad and the body thick due to the presence of an extremely large liver. Five more specimens have been caught subsequently at depths of 2012 – 2542 m.

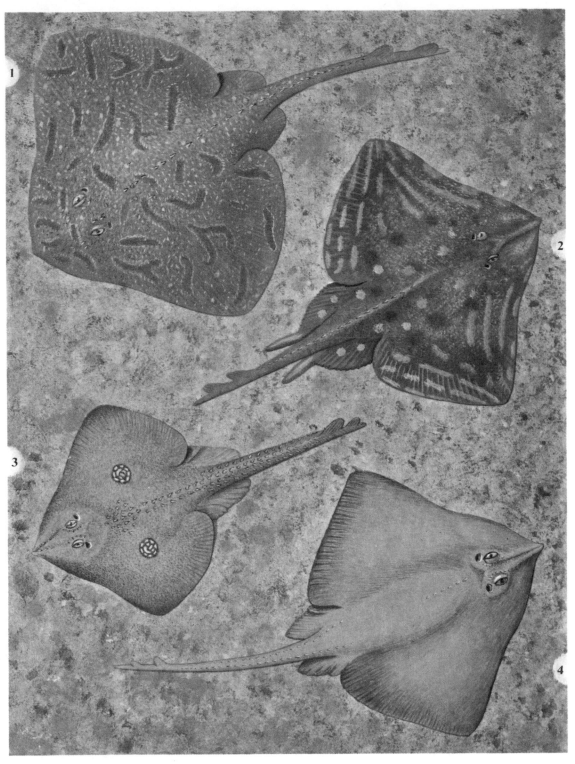

1 UNDULATE RAY
2 SMALL-EYED RAY
3 CUCKOO RAY
4 SHAGREEN RAY

RAYS AND RABBIT-FISH

Two rays (1, 3) on this page belong to the order Batoidea (p. 10) and the family Myliobatidae. This family is characterized by the well-developed pectoral fins which are free from the head. They are bottom-feeding and have a venomous tail-spine. The devil fish belongs to the family Mobulidae, the Manta rays. The rabbit-fish (2) belongs to the sub-class Holocephali, which separated from the sharks perhaps some 410 million years ago and which can be recognized by the fusion of the upper jaw to the skull, and by the single covering and external opening, on either side, for the four pairs of gills.

1 **Myliobatis aquila** (Eagle Ray) may be found all round the British Isles, generally between the surface and depths of 111 m. A number have been caught recently in Scottish waters during winter and this may indicate a migratory habit. The largest was 109 cm long and 63 cm wide. One taken in Irish waters was about 70 cm long and weighed 3·4 kg.

The mouth is on the ventral surface and the teeth form a single plate in each jaw. These provide flattened surfaces between which the food, chiefly molluscs, can be ground. Its prey is taken entirely from the sea bed and also includes crustacea. On the base of the tail, behind a small dorsal fin, is a sharp spine which injects a venom. Although much time is spent on the bottom they often swim near the surface and, seen on films taken underwater, they give the appearance of flying, their wings moving up and down in graceful motion. The eagle ray is ovoviviparous. Villi develop on the uterine wall and produce a nutritious secretion, uterine milk. This is absorbed by the embryo, at first through the filamentous external gills and later through the mouth. A gravid female of a different species, *Myliobatis freminvilli*, from the western Atlantic was found to contain six embryos folded together in pairs, the position of the head and tail being reversed in each pair. There appears to be no record of the size of the young of *M. aquila* at parturition but a small male captured in Scottish waters, was 28·2 cm wide and 49·5 cm long. The female of another species, *Aetobatus narinari*, contained a well-developed embryo 25 cm wide.

2 **Chimaera monstrosa** (Rabbit-fish, Rat-fish) is occasionally caught in deep water, where it lives on or near the bottom, at the edge of the continental slope, off Scotland and Ireland. Of those caught recently, the largest was a male 92 cm long, although a length of 150 cm can be attained, females usually being a little larger than males.

The head is large and by comparison the mouth is small. The teeth are fused to form bony plates used to grind food into fragments not more than 2 cm in length. Remains of fish, crustacea, brittle-stars, ophiuroids and polychaete worms found in the digestive tract indicate that it feeds on the seabed. It is more active at night when the large oval eyes perhaps help in the search for food.

The female is mature sexually when 78 cm long. The male (2A) has typical elasmobranch claspers, modified pelvic fins, and also a cephalic clasper. This is a bony hook with many denticles at its extremity. It may be regarded as a secondary sexual characteristic, but nothing is known of its function. The winter is spent in deep water and in spring and early summer it appears off the north and west coasts of Scotland and Ireland and in Norwegian waters. It is at this time that spawning takes place. The female lays two eggs at a time in water about 100 m deep; each egg has a chitinous capsule about 17 cm by 25 cm and is probably attached to a stone by the long tapering filament. One female, 78 cm long, captured off the Shetland Isles in April, contained two undeveloped eggs. Egg capsules containing fully developed embryos have been found off Norway. The larvae at hatching are about 10 cm in length. Young rabbit-fish have been captured off the Shetland Isles and off south-west Ireland. One specimen caught in the Bay of Biscay was 13 cm long and bore the remains of the yolk sac. From all this it seems that the rabbit-fish spawns off the British Isles and the young, after hatching, remain near the continental slope.

3 **Dasyatis pastinaca** (Sting Ray, Fire Flair) has been caught throughout the year all round the coasts of the British Isles but it is not common. The largest measured was 119 cm long, 74 cm across the wing, and gutted it weighed 18·8 kg. The largest rod-caught specimen weighed 27·760 kg.

The skin of the sting ray is smooth in the young and in most adults, though a few bear some spines in the mid-line. There is a sharp, bilaterally serrated spine on the tail, that secretes a venom. This has caused injury and even death to humans, although its effect on other marine animals is not yet clear. Molluscs, crabs and polychaetes are eaten.

The sting ray is ovoviviparous, development being similar to that of the eagle ray. The young are born in July in the Mediterranean. Some quite small sting rays, of unknown age, have been captured around the British Isles. The smallest was a male, 39·5 cm long and 24·2 cm across the disc.

Mobula mobular (Devil Fish, Horned Ray), is the largest sting ray found in European waters, although north of Biscay it is exceedingly rare. It can reach a maximum width of 450 cm. One, only 114 cm wide, was captured off the south of Ireland in 1830 and is preserved in the Royal Dublin Society Museum.

2 RABBIT FISH. Female   1 EAGLE RAY   2A RABBIT FISH. Male

3 STING RAY

THE HERRING

The class Osteichthyes includes all fish which have a skeleton of bone, a single gill-opening on either side protected by an operculum, paired fins, and skin generally covered with scales. Fertilization is external and the eggs approximately 1 mm in diameter. The super-order Clupeomorpha, a primitive group, includes the herring family Clupeidae. These are small or medium-sized, fork-tailed fish that can swim near the surface in shoals.

1 **Clupea harengus** (Herring) is one of the best known of the marine fish. Its fishery, reported from East Anglia 495 A.D., is still an important industry today. In 1968, 94·5 thousand metric tons were landed mostly in Scotland. It has a high protein content and can be used fresh, canned or preserved. Those preserved by smoking are sold as kippers, bloaters and buckling, and those by salting as roll-mops. Herrings may reach 21 to 22 years but few do so. Those caught are generally 3 – 9 years old, up to about 27 cm in length and 112 – 182 g in weight. The record rod-caught specimen weighed 368 g.

The herring begins life as an egg on the sea bottom, adhering to small stones. The eggs, about 1·2 mm in diameter, form layers 1 to 2 cm thick on large areas of the sea-bed — one was about 360 m by 3200 m.

Before hatching the embryo begins to wriggle and turn within the outer transparent egg-case. In two weeks it emerges, head first followed by the body with the yolk-sac attached. The larval fish is transparent, 5 – 9 mm long, and poorly developed; the mouth is not functional, and only pectoral fins are present, the others developing when the larva is 12 – 20 mm long. In contrast, the eyes are already prominent and well-developed indicating their subsequent importance in the life of the fish in feeding, shoaling, escape and spawning.

The larvae remain on the bottom for about 4 days and then, once the yolk has been utilized, they must feed themselves. The mouth has now developed sufficiently to take small food objects and the larvae swim upwards to the plankton where they remain for 3 to 6 months. They feed upon diatoms and flagellates, taken by darting movements made by flexing the body into an S-shape. Soon they add the nauplius larvae of crustaceans to their diet.

Metamorphosis takes place when the larva is 2·5 – 4·5 cm long, and in an aquarium was completed in 10 days, at 15°C. The body increases in depth to take on the proportions of the adult, and a coat of silver scales develops at this time. The fish now shoal and move towards the coast or an estuary, where they remain until 9 – 10 cm long. At this stage they are caught and sold as whitebait. When about 15 cm long they move to deeper waters in the centre of the North Sea.

The diet changes as the mouth increases in size. The North Sea herring larvae feed almost exclusively, for 1 – 2 months, upon the copepod *Pseudocalanus elongatus* (b). Slightly larger copepods such as *Temora* are taken as metamorphosis is reached and then larger crustacea such as *Cypris*, *Acartia*, and decapod larvae. Adult herrings selectively catch their prey, preferably *Calanus* (a) but also the arrow-worm *Sagitta* (e), the pteropod *Limacina* (g), and juvenile sand eels *Ammodytes*. If food is available, herring appear to feed throughout the year but there is a cycle of feeding and laying down of fat in readiness for the winter when fewer organisms are present. Like other members of the Clupeidae, the herring feeds, migrates and spawns in great communities or shoals. Throughout these shoals the fish tend to be regularly spaced and to move at the same speed, herring 15 – 27 cm long swimming at about 93 – 143 cm per second. New recruits remain near the top of the shoals. In the North Sea the size of a shoal may be 'that of a house' but in the northern Atlantic may be 1500 m long and some have been estimated to contain 22,000 metric tons or some 150,000,000 herrings. They spawn in September and October when the bottom temperature is between 10 and 14°C. The female stays close to the sea-bed and sheds the eggs in ribbons while males extrude milt into the sea water above the female. There is no evidence of pairing in herrings. The eggs are shed in 2 – 4 hours while the sperms remain active for three hours. A female will lay between 5,000 and 200,000 eggs per year, but few survive to become adults. The majority of the larvae hatched from eggs laid on the Dogger Bank reach the coastal waters of Denmark and the German Bight by March or April of the following year.

Besides horizontal migrations, there are daily vertical ones. Herring tend to remain in deep water or near the bottom during the day, but move towards the surface at night. Here observations from a submarine have revealed that as night falls they become motionless and remain at various angles to the horizontal 'as if drowsy'. With dawn they become increasingly active.

These vertical migrations cause fishermen to employ different methods. By day herrings are trawled from the bottom but at night they are caught in extensive nets suspended along the water from drifters which are carried by the current.

Engraulis encrasicolus (Anchovy). Shoals are found occasionally in British waters usually in estuaries or shallow bays. Spawning occurs from April to August and the eggs and fry are pelagic. A length of 22 cm can be reached.

HERRING

1 HERRING   1A young stages   1B HERRING ,eggs
a - h Herring's food, in order
a Calanus,4mm b Pseudocalanus,2mm c Meganyctiphanes,8–40mm d Nyctiphanes,17mm
e Sagitta,10–20mm   f Hyperia,15mm   g Limacina,2–5mm   h Clione,2–15mm

21

HERRING FAMILY AND DEEP-SEA FISH

The family Clupeidae (p. 20) is represented by 1 – 4, of which the shads (3, 4) enter freshwater to spawn. The order Salmoniformes (p. 88) is represented by 5, a member of the suborder Argentinoidei, elongated silvery fish with large eyes that live in moderate depths, and by 6, a member of the suborder of Stomiatoidei, deep-sea fish which possess luminescent organs.

1 **Sprattus sprattus** (Sprat) is a coastal fish inhabiting shallow water all around the British Isles. It rarely exceeds 16·5 cm, 17 g, and 5 years. It can be identified by moving a finger along the belly from tail to head as the spiny scales are rough to touch. Sexual maturity is reached between 1 and 2 years when at least 8·7 cm. Spawning occurs January to March in the Plymouth area, April to August around the Isle of Man. The eggs, 0·9 – 1·1 mm in diameter, hatch after 3 – 4 days. The larvae are 3 – 4 mm long at birth, 5 – 6 mm when the yolk has been utilized, and 32 – 41 mm at metamorphosis when the scales and adult colouration develop. Large shoals of scaled, silvery fish, 30 – 60 mm long and 6 months old, are found in rock pools mixed with young herrings. They are caught together and sold as whitebait. Until the yolk is utilized, diatoms and flagellates are eaten, supplemented by nauplius larvae and eggs until the larval fish is 8·5 mm long. Copepods form their chief food until metamorphosis after which cirripedes, copepods and larval molluscs are eaten. The adult feeds mainly on planktonic crustacea such as *Calanus*, *Pseudocalanus* and *Temora*.

2 **Sardina pilchardus** (Pilchard) is sometimes found off southern British coasts in large cigar-shaped shoals. The adults reach 24 cm and 120 g. Spawning takes place in mid-water, mainly off Cornwall. Each female sheds numerous eggs, 1·5 – 1·9 mm in diameter, that rise to join the plankton. Eggs kept in an aquarium, at 17°C, hatched in 3 days. The larva, 3 – 4 mm long at hatching, floats upside down owing to the oil globule in the yolk sac. Metamorphosis and sexual differentiation take place when the fish is 40 – 50 mm long. Larval pilchards feed largely upon copepods. At 40 – 82 mm long, the fish takes much smaller organisms while the adults feed exclusively on plankton – diatoms, peridinians, mollusc larvae, eggs, and small crustacea. They retain the small organisms because the gill-rakers end in tufts of very fine processes forming an efficient filter finer than that of the sprat, mackerel, or small herring.

3 **Alosa alosa** (Allis Shad) is a solitary fish living over the continental shelf between the surface and depths of 182 – 300 m. A length of 50 – 60 cm may be attained and a weight of 1·8 – 2·7 kg. The record rod-caught one, taken in the sea, weighed 1·488 kg. It feeds on planktonic organisms taken into the small mouth and retained by the numerous slender gill-rakers on the gill arches; 105 – 130 on the first gill arch. In spring after migrating in shoals into rivers, they spawn in mid-stream on fine gravel just above brackish water at temperatures of 17 – 24°C. The eggs, 4 mm in diameter,

slightly heavier than freshwater, and not adhesive, usually remain on the river bottom. Hatching takes 3 days at 23°C. The larvae, about 5 mm long and transparent, develop in the river and young fish migrate to the sea at the end of their second summer. Males return to the river after 2 – 5 years at sea but females usually after 4 – 5 years.

4 **Alosa fallax** (Twaite Shad) is found all around the British Isles. It enters rivers in spring and early summer to spawn. It preys upon sand eels or other fish and crustacea. A length of 50·2 cm can be attained and a weight of 964 g. One specimen, 44·5 cm long, 750 g in weight, was 9 years old. The record rod-caught specimen weighed 1·417 kg. The Killarney Shad, or Goureen, is a land-locked form found only in Killarney, where it remains without migrating to the sea. It has a deeper body and more gill rakers than the twaite shad — 43 – 53 compared with 39 – 43. It does not exceed a length of 23 cm at which size it is 5 – 6 years of age.

5 **Argentina sphyraena** (Lesser Argentine), commonly found off the Atlantic coasts of the British Isles, lives on the continental shelf, over a muddy bottom, at depths of 55 – 185 m. The larger fish inhabit the deeper water. Its diet usually consists of slow-moving, sedentary, bottom-living organisms including crustacea, lamellibranchs and polychaete worms; but in May and June planktonic species are taken and in October gadoid fish. Males reach sexual maturity when 12 cm and females when 13·5 cm long. The female first spawns when 4 years, but the male will do so when 3 years. Spawning takes place March till June. The eggs, 1·7 mm in diameter, are planktonic and are often found in April and May. Little is known of their development but post larval stages can be found at depths of 45 – 75 m. Adults can reach 21·5 cm and 70 g, females being larger than males. A male of 17 years and a female of 14 years have been found.

6 **Maurolicus muelleri** (Pearl-side) is a small deep-sea fish, up to 7·6 cm long, that has two rows of luminescent organs along the belly. It has been observed at the surface and single specimens have sometimes been washed ashore after storms. It reaches sexual maturity when about 5·7 cm long and spawning occurs in August and September. The young develop their luminescent organs when only 0·7 – 0·8 cm long, while the silvery pigment and the general adult form develop when 1·3 – 1·5 cm long. The young are usually found down to 150 m.

1 SPRAT

2 PILCHARD

3 ALLIS SHAD

4 TWAITE SHAD

5 LESSER ARGENTINE

6 PEARL-SIDE

THE MACKEREL AND FAMILY

The Perciformes (perch-like fish) is a very large order of spiny-finned fish and includes the mackerel and its relatives in the suborder Scombroidei. These fish have a row of dorsal and anal finlets. The mackerel (1), the colias (2), and the plain bonito are smaller members in which the dorsal fins are widely separated.

1 **Scomber scombrus** (Mackerel) is found all around the British Isles. To illustrate its life history the movements of a population, found around the western end of the English Channel, will be followed.

The main shoals consist of sexually mature fish, more than two years old. In October and November mackerel disappear from the surface waters, moving in shoals to the sea floor where they form dense, localized concentrations distributed over wide areas. At this time they feed upon bottom-living animals such as shrimps, amphipods, mysids, polychaete worms, and small fish. From late December until February they move outwards from their winter quarters and begin to ascend to surface waters, rising to the surface at night and returning to the bottom during the day.

The shoals migrate from January to July to the spawning grounds off the south of Ireland where the water is over 185 m deep. During this period food is taken voraciously from the plankton, including copepods, for example *Calanus* and *Anomalocera*.

The pearl-side, *Maurolicus muelleri*, is also taken – 13 were found in one mackerel's stomach. Underwater observations have shown that the planktonic prey is attacked selectively.

The onset of sexual maturity occurs in mackerel 29·3 cm long, and almost two years old. These fish when fully mature join the main shoals which migrate to the spawning grounds. In 1868 Sars wrote of the spawning mackerel that it 'lays its eggs at some leagues from the shore and at the very surface of the waves, where a great quantity of these fishes may often be met with engaged in spawning'. A medium sized female may produce a total of 360,000 to 450,000 eggs, 1·0 – 1·4 mm diameter, during a spawning season. She sheds only a few of them at a time – 40,000 to 50,000 — as they ripen.

Hatching takes 9 days at about 10°C. The developing eggs and young larvae, both planktonic, are carried by the currents towards the English Channel and the Irish Sea. Larvae 4 to 15 mm long have been captured during July and August in water deeper than 120 m. As the larvae grow they move towards the shore — but information about this phase of the life cycle is sparse. The food of larvae 5 – 13 mm long consists chiefly of the nauplius larvae and eggs of copepods. Small fish are also taken by larvae more than 9 mm long. The mackerel grows very rapidly in its first year, to 20 cm or more and 58 g. After 7 years it is 36 cm long and 350 g; a fish of 20 years is 43 cm and weighs 669 g. Occasionally a very large mackerel is captured and one fish, 46·1 cm long and 851 g in weight, was estimated to be 30 years of age; another one 60 cm long weighed 1·74 kg. The record rod-caught mackerel, weighing 2·452 kg, was caught near the Eddystone Lighthouse.

2 **Scomber colias** (Colias, Spanish Mackerel) closely resembles the mackerel but has a larger eye and a corselet of large scales in the pectoral region. Three specimens caught recently off the Irish coast weighed 343 – 570 g and were 34·1 – 39·6 cm long. Found in the Mediterranean and both sides of the Atlantic, it is an irregular summer visitor to British waters.

Auxis thazard (Plain Bonito, Frigate Mackerel) has a cosmopolitan distribution in tropical seas. It has been caught on rare occasions to the south of the British Isles, the largest being 42 cm and 1·15 kg. It can reach a length of 61 cm and weigh 2·72 kg. It resembles the mackerel but is stouter, and can be distinguished from other members of the family by the two widely separated dorsal fins and a corselet of large scales in the pectoral region.

Argentina silus (Great Silver Smelt, Greater Argentine) is a deep-water species of the order Salmoniformes (p. 88) and the suborder Argentinoidei (p. 22) that has been caught at a depth of 914 m but more usually between 90 and 370 m. It is found off the west and north of Scotland and Ireland. It is an elongated fish, yellowish in colour with an underlying silvery sheen, which has 67 lateral line scales, a larger number than *A. sphyraena* (p. 22). It can reach a length of 55 cm but the largest one caught recently around the British Isles measured 44 cm. A smaller specimen of 34·6 cm weighed 340 g. Spawning takes place from May to September. The eggs, 3 to 3·5 mm in diameter, are bathypelagic and the larvae when newly hatched are colourless, 6 to 9 mm long. The eggs and larvae have been found at 900 m but are more usually taken at depths of 300 to 400 m.

1 MACKEREL   2 COLIAS MACKEREL

TUNNIES AND BONITOS

The larger members of the suborder Scombroidei (p. 24) illustrated on this page all have the dorsal fins close together and a conspicuous lateral keel just in front of the tail.

1 **Thunnus (Germo) alalunga** (Long-finned Tunny, Albacore). This rare visitor, essentially a warm water species, is near the northern limits of its range in British waters, where it is usually found at the edge of the continental shelf. It is distinguished from other species of northern Europe by the very long pectoral fins. Spawning takes place in the Mediterranean between July and September. The eggs, 0·84 – 0·94 mm in diameter, are pelagic. Growth is rapid after hatching; at 1 year it is 17 – 18 cm long, at 7 years 91 – 93 cm. One specimen caught in Morecombe Bay was 80 cm long and weighed 10 kg; the pectoral fin was 30 cm. A predator, it feeds upon anchovies, skippers, lantern and hatchet fish, as well as cephalopods and crustacea.

2 **Thunnus thynnus** (Tunny, Bluefin Tuna). This oceanic, off-shore fish can be seen in large numbers from July to October to the north of the British Isles and in the North Sea. It was first reported by herring fishermen in the late 1920s, and now there is an annual meeting of game fishermen at Scarborough and Whitby where, in 1933, the record rod-caught specimen of 385·989 kg was captured. Its body is streamlined and while it swims the dorsal fin is retracted in a groove, and the pectoral and pelvic ones are folded into hollows against the side of the body, thus reducing turbulence. This tunny has a swimming speed of 1·75 m per second and can swim for great distances. Tagged specimens have averaged at least 65 km per day for 119 days.

The home of the tunny is off the west coast of Portugal and Spain and in the Mediterranean. During April they collect together into schools, in preparation for spawning in June and July near the Azores, near Gibraltar, and between Sicily and Sardinia. These schooling fish do not feed or take bait. Development of the gonads is rapid, probably taking only a few weeks prior to spawning. Observation of tunny while spawning revealed that two fish would rise to a depth of 8 – 10 m and roll around. Their ventral surfaces touched together and the eggs and milt were released. The fish then descended. This procedure was repeated several times. On completion of spawning the tunny travelled northwards on feeding migrations. Some penetrated the North Sea where they preyed upon the herring shoals. They also eat other plankton-eating fish including mackerel and sardines. Gadoids, as well as squids and crustacea, are also eaten. The eggs, 1·0 – 1·1 mm in diameter, hatch after two days into larvae 3 mm long. Young fish hatched in June will be about 400 g by September, and a year later 4·5 kg. By 3 years they are 16 kg, nearly 100 cm long,

sexually mature, and ready to spawn. A length of nearly 400 cm and a weight of 700 kg may be attained; in British waters specimens of up to 274 cm have been captured. An age of 14 years may be reached.

The tunny is one of the few fish that maintain their body temperature about 3°C above that of the surrounding water. This was first demonstrated by Sir Humphrey Davy. This feature is presumably related to its large size and considerable activity. The oily, red flesh is very suitable for canning, and a large tunny fishing industry has developed. In 1968 the world catch was 101,000 metric tons.

3 **Katsuwonus pelamis** (Oceanic Bonito) is the only tunny, occasionally found in British waters, that has 4 – 6 parallel dark blue stripes on the otherwise white belly. It usually remains in the open sea where the temperature is 18 – 20°C. In tropical waters it spawns, from April to September, in coastal areas and the open sea. It does not spawn in northern waters. Growth is rapid and it is 45 cm long at 1 year, and 72 cm at 2 years when sexual maturity is reached. A school of feeding oceanic bonitos was observed in Hawaiian waters. The 100 or so fish dived simultaneously and were away from the surface from 3 seconds to 28 minutes. The movements were slow, abrupt and vertical, as the fish vanished from view. The small fish taken during these dives were *Synodus variegatus*, with distinctive colour pattern and *Holocentrus lacteoguttatus*. They also prey upon other fish such as mackerel, clupeids, and lantern fish as well as squid and crustacea.

4 **Sarda sarda** (Pelamid, Short-finned Tunny, Belted Bonito) is widely distributed in the warmer regions of the open Atlantic. Although only rarely caught around the British Isles, it has been recorded from the east coast of Scotland, the western part of the English Channel, and from Ireland where it is common in the waters well off the south coast. It can reach a length of 91 cm but the largest caught in British waters was 76·7 cm and 5·7 kg. The record rod-caught specimen weighed 4·004 kg.

Maturity is reached when about 2 years and spawning occurs in May and June in the Mediterranean. The eggs, 1·2 to 1·32 mm in diameter, are pelagic.

Euthynnus alletteratus (Marbled Tunny). One specimen was caught in July 1952, off the west coast of Scotland at Garlieston, Wigtownshire. It was 61 cm long and a cast of it was made in the Royal Scottish Museum.

1 LONG-FINNED TUNNY   2 TUNNY
3 OCEANIC BONITO   4 PELAMID

27

SEA BREAM

The fish on this page belong to the order Perciformes, (p. 24) and its main suborder, Percoidei, which contains many typical perciform fish. The sea breams are all included in the family Sparidae, fish whose deep body and head is covered with hard scales. The teeth are specialized and show differentiation into chopping, incisor-like front teeth and rounded, crushing, molar-like teeth.

1 **Pagellus bogaraveo (centrodontus)** (Red or Common Sea Bream) is a fish of the eastern Atlantic that approaches the coast of Ireland from deeper water, will enter the North Sea, and also appears off Norway. During the annual inshore migration the fish segregate according to size. The larger and older fish of 22 to 46 cm are found in deeper water, down to 200 m, on the continental shelf; those of 22 to 31 cm are at 50 m depth. A length of 50 cm may be attained. The record rod-caught specimen weighed 3·402 kg.

Adult fish more than 20 cm in length have a large diffuse black spot at the origin of the lateral line, but this is absent in smaller fish. The mouth is large and the teeth in the front of the jaw are short, curved and sharply pointed. Along the sides of the jaw there are numerous small, rounded, molar teeth of which there are 2 or 3 rows in the upper jaw and 2 rows in the lower one. The diet is known to include decapod crustaceans, echinoderms and fish, particularly young hake.

In the summer and autumn spawning takes place, mainly in water of more than 100 m depth. The eggs are planktonic. It has been suggested that this fish is slow in reaching sexual maturity and that one of about 44 cm length is 17 years old.

The red sea bream is fished for commercially along the continental shelf off the coast of Ireland. It has formed part of man's diet since Neolothic times, some 4000 years ago, bones of *P. bogaraveo* having been found on Shell Mound on the island of Oronsay, off the west coast of Scotland.

2 **Pagellus erythrinus** (Pandora), is a Mediterranean species, a rare wanderer that occurs off the southern and western coasts of the British Isles and even enters the North Sea.

The eye is much smaller than that of the red bream as is the mouth, and the lateral line has 50 – 60 scales. It feeds mainly on fish and crustacea as well as taking annelids and lamellibranchs. The pandora spawns in spring and early summer off the Algerian coast, but it is not known to spawn in northern waters. The eggs are pelagic. It may reach a length of 35 cm.

3 **Spondyliosoma cantharus** (Black Bream, Old-wife) is found off the western shores of the British Isles, in the English Channel, and occasionally in the northern North Sea. It has a local distribution and when inshore it inhabits rock faces. A length of 50 cm can be attained but the usual length is about 30 cm; a black bream reported recently from Irish waters was 33·4 cm long

and weighed 539 g. Compared with the red bream the eye of the black bream is larger, the body more compressed and the pectoral fin relatively shorter. There are about 72 scales in the lateral line. The teeth of the black bream are conical and although little is known of its feeding habits it appears to take algae and rock encrusting animals.

In May, in British waters and in the Mediterranean, the black bream comes inshore to spawn. The male makes a depression in the sand with his tail and on this the female lays a single layer of adhesive eggs, each about 1 mm in diameter. These are guarded and aerated by the male and hatch in 9 days at 13°C.

Although edible, the black bream is rarely eaten but it provides some sport for anglers. The record rod-caught specimen weighed 2·749 kg.

4 **Sparus aurata** (Gilt-head) is a very rare visitor to the British Isles. Most records from northern Europe are from the last century, but in 1970 a record one of 680 g was taken on a rod.

5 **Boops boops** (Bogue) is a rare vistor that has been caught with some regularity in the past ten years off the south and south-west coasts of the British Isles. It can reach a length of 40 cm and one specimen caught recently in Irish waters was 36 cm long and weighed 500 g. The record rod-caught specimen weighed 733 g.

Not typical of the sea breams in shape the bogue has a long shallow body but the dentition is characteristic with notched incisor-like teeth in each jaw. It appears to feed mainly upon algae, sponges and encrusting animals.

In the Mediterranean spawning takes place from April until July. The eggs are pelagic and about 0·89 mm in diameter. The fry are 2·5 to 3·2 mm long at hatching. The bogue is not known to spawn in northern waters.

6 **Pagrus pagrus** (Couch's Sea Bream) is only known in the British Isles from a specimen caught off Polperro, Cornwall on 8 November in 1842. It was 50 cm long and weighed 2·7 kg.

Dentex dentex (Dentex) lives near rocks at depths of 10 – 200 m but may go into deeper water in winter. The largest of our sea bream, it can reach a length of 142 cm and a weight of 12·7 kg. It has large canine-like teeth at the front of each jaw, succeeded by many smaller teeth. An active predator, it takes cephalopods and fish.

1 RED SEA BREAM   2 PANDORA
3 BLACK BREAM   4 GILT-HEAD
5 BOGUE   6 COUCH'S SEA BREAM

PERCH·LIKE FISH

All the fish on this page belong to the suborder Percoidei (p. 28). The family Carangidae, represented by 1 – 3, is widely distributed and includes tropical, pelagic fish that exhibit summer inshore migratory movements. The family Cepolidae is represented by 5, a Mediterranean species that enters the Atlantic and migrates inshore to spawn. Fish of the family Mullidae, represented by 4, have a deep head and two long barbels beneath the jaw. Ray's bream belongs to the family Bramidae, relatively large oceanic fish with deep, compressed bodies.

1 **Naucrates ductor** (Pilot Fish) is found in warm seas but has been reported off south and south-west England and off Scotland. A length of 36 cm can be attained; the largest taken in British waters was 31·6 cm long and weighed 332 g. A boldly marked fish, it is noted for its habit of accompanying sharks, turtles, and even ships. The *Kon-Tiki* had 60 – 70 pilot fish around it after their real hosts, the sharks, had been killed by the crew. Scraps of food left by the sharks as well as its parasites are probably eaten by the pilot fish. It is believed to spawn in the open sea late in summer.

2 **Trachurus trachurus** (Scad, Horse Mackerel) is common around the British Isles with the exception of the North Sea north of Norfolk. It can reach a length of about 30 cm and the weight of the record rod-caught specimen was 1·488 kg.
It resembles the mackerel but the body is deeper and the head and eye much larger. The lateral line has a deep curve along which are about 80 conspicuous scales or scutes. At dusk it comes near the surface to pursue fry, especially those of herring and sprats; these and small fish are its main food, but euphausids are also taken. It is sexually mature at 14 cm long. Spawning occurs around the British Isles April — August. The egg, 0·95 mm in diameter, has a fairly large oil-globule. Newly hatched larvae are 2·5 mm long and the yolk sac projects beyond the head. A long period of pelagic life ensues. Larvae 1 – 7 cm are often found beneath the umbrella of jellyfish which provides protection from predators. Larvae hatched early in the year are 7 – 10 cm long by September and a year later are 17 – 20 cm.

3 **Seriola dumerili** (Amberjack) was added to the British fauna after a specimen 45 cm long and 1·020 kg in weight was caught in the Salcombe estuary, Devon, in 1951. An oceanic species found in the warm parts of the Atlantic, it spawns in summer off the coast of North Carolina but its eggs are unknown.

4 **Mullus surmuletus** (Red Mullet) is a fairly regular visitor to the south and south-west of the British Isles and in the North Sea off Scotland. Specimens of up to 40 cm length have been captured. The record rod-caught specimen weighed 1·644 kg. Food consists of shrimps, amphipods and other small crustacea as well as polychaetes, small bivalves, gastropods and cephalopods. Its barbels are used to rake the sea-bottom in search of food hidden in the sand. It was much prized as food in ancient Rome and was kept alive in tanks in their dining-rooms until selected to be eaten.
Sexual maturity is reached when about 14 – 15 cm long, usually when just over a year old. Breeding fish have been caught as early as April in the North Sea where there are records of it spawning in June at depths of 25 – 40 m. The eggs are 0·85 – 1·0 mm in diameter and pelagic. Hatching takes place in a few days and the larvae are about 2·8 mm long. They are distinguished by the attached yolk projecting in front of the mouth, much as in the scad.

5 **Cepola rubescens** (Red Band-fish) is a species of the Mediterranean and neighbouring Atlantic. It generally arrives early in the year in the English Channel, Irish waters, and occasionally off the Scottish coasts and the Isle of Man. The migration into relatively shallow water may be connected with spawning which occurs from June to August in the English Channel. The eggs, probably pelagic, are about 0·72 mm in diameter, transparent, and have an oil globule. A post larva 7 mm long was taken in September off Plymouth.
Specimens of 50 cm have been captured in British waters. Apart from distinctive shape and colouring the red band-fish has two curious features, its vertical swimming position and the habit of sucking sand into its mouth to form a burrow.

Brama brama (Ray's Bream) is a deep-bodied fish with a steep profile. The back is a dark green-brown, the sides are a brilliant silver, the dorsal and anal fins have dark edges, and the pectoral fins are yellow. Specimens captured in British waters have been up to 65 cm in length. The weight of the record rod-caught one was 3·621 kg.
It is believed to swim in deep water in the Mediterranean and Atlantic in winter coming to the surface in summer at which time it may migrate northwards. Ray's bream has been found in many places off the British Isles and even penetrates the North Sea. It has been captured alive most frequently between September and October, but during November and December as the temperature falls a number have been stranded on the coast. Often several are captured in the same vicinity. Spawning occurs in mid-Atlantic south of the equator; young specimens 1·6 cm and 5·3 cm long have been obtained drifting northwards.

1 PILOT FISH

2 SCAD

5 RED BAND-FISH

3 AMBERJACK

4 RED MULLET

GARFISH, SKIPPER, AND FLYING FISH

The fish on this page represent three different families in the order Atheriniformes. Fish of this order have soft fin rays, pelvic fins in an abdominal position, and a very low lateral line close to the belly. They are pelagic fish whose egg capsules are provided with filaments.

1 Exocoetus volitans (Flying Fish) is found in the Atlantic and is occasionally carried to the south-west coasts of the British Isles.

The 'flight' of this fish is probably an escape response. Flight is achieved by the fish approaching the surface with pectoral and pelvic fins folded into the body so that a high initial speed can be attained before leaving the water. Once the body of the fish has left the water, the flight is accelerated by powerful strokes of the tail and the fish is able to soar into the air supported by the outspread pectoral fins. The flight continues without movement of the pectoral fins. Large fish are able to cover more than 30 m.

The eggs of the flying fish are provided with filaments and may be pelagic.

2 Scomberesox saurus (Skipper, Saury Pike) is a pelagic fish of the Mediterranean and Atlantic that migrates in shoals to the continental shelf in the summer. It appears sporadically off the warmer Atlantic shores of the British Isles and even in Scottish waters.

In general form it resembles *Belone* but is smaller, not usually exceeding a length of 45 cm. The body is less elongated, the bill relatively shorter, and the teeth finer and smaller.

A surface living fish, it has the habit of leaping into the air. It often skims over the water for considerable distances, apparently on the tips of its pectoral fins, being propelled by violent oscillations of the lower lobe of the tail.

The skipper moves away from the coast to spawn in mid-ocean. The eggs, although provided with filaments, are pelagic as also are the larvae. These have been obtained in great numbers in the tropical Atlantic, and are carried by the currents towards the north Atlantic. The young undergo metamorphosis much as *Belone* (3) does, the lower jaw growing forward first.

3 Belone bellone (Garfish, Greenbone) is found in the Black Sea, the Mediterranean, the eastern Atlantic and north to Norway. A pelagic fish, it arrives off the coast of the British Isles in April or May some weeks before the mackerel, and from this habit it is sometimes called the 'mackerel guide'.

It is a long slender fish whose jaws, elongated into a snipe-like bill, carry many teeth. It feeds upon amphipods, euphausids and annelids but its feeding habits have been little studied.

Shortly after the inshore migration, spawning takes place in shallow, weedy water in May and July. The eggs are large, 3 – 3·5 mm in diameter, and have long filaments which entangle the eggs with each other and with the weeds. Incubation lasts about 36 days and at hatching the larvae are 13 mm long, the yolk having been almost completely utilized. At this time the jaw has only a slight prominence and not until the larva is 50 mm long does it begin to elongate. By September the young garfish are 65 – 75 mm long, by the following summer 15 – 17 cm, and at the end of two years 21 – 22 cm long. A length of 82 cm may be attained but the usual length is 60 to 70 cm. The record rod-caught specimen weighed 1165 g.

The garfish, while good to eat, is also interesting to have on the table as its bones are translucent and green in colour.

1 FLYING FISH
2 SKIPPER
3 GARFISH

33

COD AND FAMILY

The fish on this page are all members of the order Gadiformes and the family Gadidae (p. 36).

1 **Melanogrammus aeglefinus** (Haddock) is found all around the British Isles but is fished commercially mainly in the North Sea.

The haddock reaches sexual maturity when 2 – 3 years of age and 22 – 30 cm in length. At first spawning a fish 80 g in weight and 22 cm long may produce about 12,000 eggs while an older fish of 6·9 kg and 85 cm may lay as many as 3 million. Spawning takes place near the sea-bed but the eggs, 1·2 – 1·6 mm in diameter and without an oil globule, are pelagic. They hatch in 13 days at 8°C, and the larvae are pelagic. They feed upon crustacea and during the first year of life become demersal, subsequently feeding upon bottom-living organisms including echinoderms, worms, crustacea, molluscs and fish. By the end of 8 years a haddock is about 60 cm long; although older fish are rarely captured, they can survive in captivity for 14 – 15 years. The record rod-caught specimen weighed 4·209 kg.

2 **Merlangius (Odontogadus) merlangus** (Whiting) is found all around the British Isles, and is fished commercially in the English Channel, Bristol Channel, Irish Sea and North Sea.

It spawns between January and September at depths of 20 – 60 m. The eggs are 0·97 – 1·32 mm in diameter. At hatching the larvae are 3·2 – 3·5 mm long and they drift towards the coast in association with jellyfish. As the larvae grow, they gradually move downwards to reach the bottom when 5 – 15 cm long. They are 24 cm in length at 2 years, 32 cm at 3 years, and 50 cm and about 950 g at 6 years after which their length increases little although they may live for 8 years. The record rod-caught specimen weighed 2·811 kg.

Their diet consits mainly of fish such as sand eels and smaller whiting and pollack, as well as crustacea. The whiting, unlike the haddock, is not primarily a bottom-feeder.

3 **Gadus morhua** (Cod) is found all around the British Isles and is commercially the most important of the demersal fish. Cod shoal where the bottom is of sand or mud, congregating where the food is abundant.

The egg is almost spherical, about 1 mm in diameter, and the yolk transparent. It is pelagic and, at 6°C, takes about 2 weeks to hatch. The young emerges tail first from the egg and then drifts out to sea, where it is swept into the main current flowing north. A current of some 10 cm per second could carry a young fish about 965 km in 5 months. The young cod travels in midwater to nursery grounds where it remains for 3 – 6 months and is able to feed on small herrings. When 2 cm long it moves to the bottom at depths of 200 – 400 m.

Sexual maturity is reached when a length of 80 cm or more is attained, the males maturing slightly earlier than the females. First spawning occurs at 6 – 14 years of age but the majority are 8 – 10 years. Mature cod return from the feeding grounds, some 800 – 1300 km distance, to the spawning grounds where the larger and older females arrive first, followed by the males. The spawning areas of the Norwegian fjords have a layer of cold water on top of the warm Atlantic water and the fish gather at the boundary of these two layers of water. Spawning takes place in mid-water or close to the bottom depending upon the temperature.

Spawning behaviour of the cod, observed in aquaria, is complex. The male shows territorial and aggressive behaviour and is spatially separated from the rest of the group. In display the median fins are erected and the body undergoes exaggerated lateral movements. Grunts are emitted, the sound being produced by the swim-bladder, and are accompanied by jerky movements. These displays excite the female and, if ready to spawn, she will follow the male. The male will mount the female and then slip down one side until inverted and below her, their ventral surfaces and genital apertures closely pressed together. They spawn while swimming in circles which may help to mix the eggs and milt as they are extruded.

Each female may produce 500,000 – 15,000,000 eggs depending upon her size. These are laid over several days, for as the eggs swell with fluid just prior to extrusion, the body would be unable to hold them all if they became swollen at the same time. After spawning, the spent cod are carried away from the Norwegian fjords by the same currents that take the baby cod to the coastal banks of the Barents Sea. The adults thus travel some 1600 to 2600 km between spawning and feeding grounds.

The young cod prefer to feed upon small crustacea but as they get older and are able to swim more rapidly, they take larger food and a greater proportion of fish. Detection may be by sight as the cod can take moving objects 2 mm in size; taste-buds on the barbels and the pelvic fins enable them to follow odours from food objects. Coelenterates, worms, crustacea, molluscs, echinoderms are all eaten as well as fish which in a large cod, 70 – 100 cm in length, may form 70 per cent of the diet.

At hatching the cod is 0·7 cm long and when 1 year is about 18 cm long and weighs 400 g. By 6 years it is 89 cm in length and weighs 7 kg. Cod can live as long as 20 years and reach considerable size. The record rod-caught specimen weighed 20·878 kg.

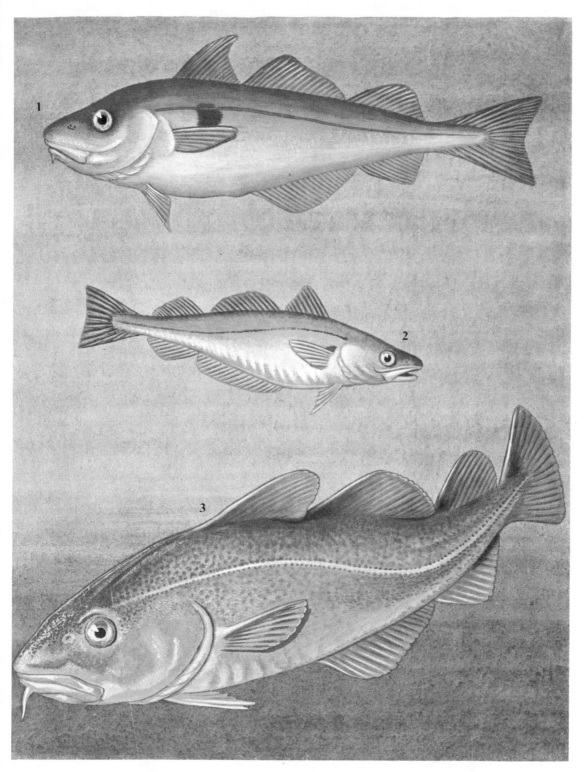

1 HADDOCK
2 WHITING
3 COD

COD FAMILY

Fish of the order Gadiformes have soft fin rays, generally very long, many-rayed dorsal and anal fins, pectoral and pelvic fins close together, and tail fin reduced or absent in some members. This order includes the cod family, Gadidae, which contains many species, mostly marine but with a freshwater member found in British rivers, *Lota lota* (p. 92). Fish of this family have 1 – 3 dorsal fins extending along most of the back, 1 – 2 anal fins extending from vent to tail, and pelvic fins anterior to the pectoral ones. The skin is covered with small scales.

1 **Trisopterus (Gadus) luscus** (Pouting or Bib) is common around the British Isles but not commercially fished. It has a very deep body and rarely exceeds a length of 30 cm. The record rod-caught specimen weighed 2·494 kg. Sexual maturity is reached in the male when 13 cm and in the female when 18 cm long. Spawning takes place from January to June at depths of 50 – 100 m. The eggs, 1·0 – 1·2 mm, are pelagic as also are the larvae, 3 mm long at hatching. They drift shorewards, often in association with large jellyfish, feeding upon copepods, usually *Calanus*, until about 45 mm long. When 70 mm long the young fish are demersal and may be found in sandy bays where they feed upon small crustacea. Later when 135 – 450 mm long they move offshore to muddy sand or mud where they feed on small crustacea including decapods, amphipods, and mysids as well as cephalopods and other small molluscs.

2 **Trisopterus (Gadus) minutus** (Poor Cod) is the smallest member of the cod family in British waters, where it is common. It can reach a length of 28 cm but the majority are less than 20 cm. The record rod-caught specimen weighed 283 g. The male and female reach sexual maturity when 11 and 13 cm respectively. Spawning takes place March till June on coastal banks at depths of 100 – 200 m. The eggs are 0·95 – 1·07 mm in diameter and the fry at hatching are 2·75 mm long. They drift to the coast, rapidly leaving the surface waters to arrive inshore on the bottom when 15 – 25 mm long. Here they feed mainly on amphipods, copepods and isopods. Poor cod 5 years of age have been recorded.

3 **Trisopterus (Gadus) esmarkii** (Norway Pout) is found off the north and west coasts of the British Isles and in the North Sea. It can attain a length of 24 cm.
Sexual maturity is reached after one year and spawning occurs January till April. The eggs, 1 – 1·08 mm in diameter, are pelagic and the young at hatching are 3·2 mm long. They drift with the current feeding upon copepods, euphausids and arrow worms. Young fish reach the bottom in summer when 6 – 7 cm long and have been found off the Faroes, Scotland and in the northern North Sea. They feed on small crustacea, such as amphipods and shrimps as well as small fish. At 1 year a length of 11·5 cm is reached, at 2 years 16 cm and at 3 years 20 cm.

4 **Phycis blennoides** (Greater Fork-beard or Forked Hake) is a deep water Atlantic species which migrates in winter to the English Channel, Irish and Scottish waters, and the North Sea. It can attain a length of 76 cm by about 20 years of age. One caught recently was 46·4 cm and 980 g. The record rod-caught one weighed 2·133 kg. Spawning takes place in early summer in oceanic waters and the eggs and larvae drift to coastal waters. The spent adults also migrate into shallow water to recuperate.

5 **Raniceps raninus** (Lesser Fork-beard or Tadpole-fish) is found around the British Isles off rocky coasts but is not very common. Specimens 10 to 26 cm long have been captured. The record rod-caught specimen weighed 403 g. Spawning takes place in summer and autumn around the coast in shallow water. The eggs are 0·8 mm in diameter. Newly hatched larvae, 2·5 mm long, have very long pelvic fins. Pelagic larvae have been taken in the Moray Firth in October and off the south-west coast of Ireland in August and September.

Brosme brosme (Torsk) is found off the north and west coasts of the British Isles and sometimes in the North Sea, usually at depths of 100 – 1000 m. Unlike most of the gadoids, it does not tend to shoal. Blue and yellow above with pale grey belly it may be 105 cm long. The record rod-caught specimen weighed 5·471 kg. The dorsal, anal and tail fins have white edges and a dark submarginal band; the long single dorsal and anal fins are joined at their bases to the caudal fin. While the body is elongated, and covered with very small scales, the head is relatively large, and the diameter of the eye equals the length of the barbel. The lateral line begins high above the pectoral fin and bends downward at the level of the vent. Spawning takes place from April to June and the eggs, 1·3 – 1·5 mm in diameter, have an oil-globule. Hatching occurs in 9 days and the young fish have elongated pelvic fins. The young stages are found in relatively shallow water, 10 – 15 m deep.

Micromesistius (Gadus) poutassou (Poutassou or Blue Whiting) is a pelagic, shoaling, off-shore species common in Irish waters at depths of 300 – 400 m. It has also been found in the English Channel and the Firth of Clyde. It can be 41 cm long. The record rod-caught specimen weighed 267 g. Although not fished commercially it is the prey of ling, hake and cod.

1 POUTING

2 POOR COD   3 NORWAY POUT
4 GREATER FORKBEARD   5 LESSER FORKBEARD

COD AND HAKE FAMILIES

The fish on this page belong to the order Gadiformes (p. 36); 2–4 belong to the cod family, Gadidae (p. 36), and 1 to the hake family, Merlucciidae, whose members have, in both jaws, long, sharp teeth directed backwards, and hinged to allow large pieces of food to enter.

1 **Merluccius merluccius** (Hake) is found all around the British Isles except in the central and eastern parts of the English Channel. Its range extends from Norway to Morocco.

Some males reach sexual maturity when only 28 cm long, at 3 to 6 years of age. The females, however, are 65 – 75 cm long and 6 – 8 years of age at maturity, and usually 8 years when they first spawn. There is a general shoreward breeding migration during the spawning season which lasts from April to October. Fish that spawn early do so in deep water beyond the 600 m line. As the season proceeds, so the fish move progressively into shallower water to spawn. In one breeding season a female may deposit between 500,000 and 2,000,000 eggs. At the end of the season the hake return to deep water where the larger fish live in deeper water than the smaller ones.

The eggs, spawned mainly in waters to the south of Ireland, are about 1 mm in diameter. The eggs and the larvae, which are 3 mm long when they hatch in about 7 days, are carried eastwards in the plankton, by the drift of the surface water, to shallower coastal areas. Little is known of the first 2 years of life except that the young hake is pelagic. At 1 year it is 10 cm long and at 2 years 20 cm and 50 g in weight, by which time it is able to swim strongly.

After 2 years of age the hake then increases by 8 – 9 cm per year until 8 years, but after 3 years the male grows a little less rapidly than the female. After 8 years males and females show smaller increments in length. The average length of a hake of 11 years is 85 cm and its weight about 4 kg. A length of 100 cm is seldom exceeded. The lifespan is not known as the rings on the otolith, or ear-stones, that are used to estimate age, cannot be analysed in fish of over 12 years. The record rod-caught hake weighed 11·495 kg.

The feeding habits pass through several stages. When newly hatched, it eats nauplius larvae and small copepods. During the early pelagic phase the young hake eats krill, chiefly euphausids and amphipods. At about 3 years of age it becomes almost entirely fish-eating. The chief prey is *Micromesistius poutassou* when the hake is living off shore but smaller hake are taken as well as other mid-water fish.

2 **Molva molva** (Ling) is found all around the British Isles although less frequently off the southern coasts. It is a bottom-living fish with a preference for rocky ground, in water of 10 – 300 m, most usually 40 – 100 m. It is a valuable commercial fish caught by long-liners. A length of 150 – 180 cm can be reached and a weight of 32 kg. The record rod-caught specimen weighed 20·410 kg.

Sexual maturity is reached in the male when 80 cm long, and in the female when 90 – 100 cm long and 6 – 8 years of age. Spawning occurs in depths of 100 – 200 m from April to June. The pelagic eggs are 1·0 – 1·14 mm in diameter. Females of 8 and 24 kg can produce 12 and 28 million eggs respectively. At hatching the larva is about 3·2 mm in length and a long period is spent in the open sea before reaching the bottom when about 7 cm long. Ling feeds mainly on fish, including mackerel, megrim, dabs and scad; cephalopods are also sometimes taken.

3 **Pollachius (Gadus) pollachius** (Pollack) is found all around the British Isles although not so frequently off the northern shores. During the summer it is found on rocky bottoms at depths of 2 – 40 m; some of the largest have been seen close to wrecked ships. A length of 98 cm may be reached when 10 – 11 years of age. The record rod-caught specimen weighed 10·658 kg.

Spawning takes place on coastal banks from February to May, at depths of 100 m, and afterwards the spent fish often congregate over shallow shelves to feed. The eggs are 1·1 – 1·22 mm in diameter and the newly hatched larvae are pelagic. They feed upon copepods and euphausid larvae. When 12 – 42 cm long they are demersal and live offshore over rocky outcrops. As they grow larger they live over muddy sand and their diet includes cephalopods and small fish.

4 **Pollachius (Gadus) virens** (Coalfish, Coley) is found all around the British Isles although with less frequency off the south coasts. It may reach a length of 120 cm and a weight of 16 kg, and one in captivity reached an age of 10 years. The record rod-caught specimen weighed 11·850 kg. Sexual maturity is reached at a length of 58 cm when 4 years of age. Spawning takes place over a hard bottom at depths of 100 – 200 m from January to April. The spent fish afterwards gather over submarine shelves to recover. The eggs, 1·02 – 1·28 mm, are pelagic and when newly hatched the larvae are 3·6 mm long. They remain in the plankton until about 53 mm long feeding on copepods, euphausid larvae and fish eggs. When 70 mm long they become bottom-living and are found in bays feeding upon isopods and amphipods. As they grow larger, 40 cm or more, their diet includes young gadoids, sand eels, smaller coalfish and cephalopods.

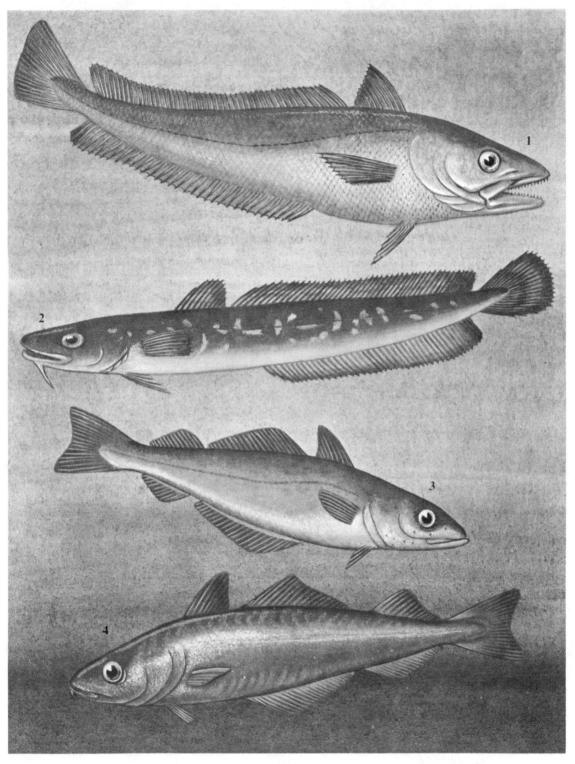

1 HAKE
3 POLLACK

2 LING
4 COALFISH

DEEP WATER AND OCEANIC FISH

The fish on this page belong to four orders. The order Lophiiformes (5) is characterized by a dorsal fin modified to form a lure and pectoral ones used in movement along the bottom. The order Zeiformes (4) contains deep-bodied fish that have thick scales and are pelagic, being found often in deep water. The order Scorpaeniformes (1, 2), mail-cheeked fish, sometimes called oceanic perches, have well-developed bony armature and produce live young, an almost unique feature amongst bony fish. The opah (3) belongs to the order Lampridiformes and lives in the upper waters of the open sea.

1 **Sebastes marinus (norvegicus)** (Norway Haddock or Redfish) is occasionally found to the west of Scotland. A fish of the open ocean, it lives over deep water at depths of 200 – 400 m. It can reach a length of 100 cm. *S. marinus* and *S. viviparous* (2) are ovoviviparous, and are amongst the few bony fish where fertilization is internal. Up to 3500 young develop within the mother and are liberated in oceanic waters in May or June when about 5·5 mm long. The young fish are carried to the continental shelf and into shallow water. In Dingle Bay, Ireland, October 1968, a small specimen was captured 14 cm long and weighing 34 g.

2 **Sebastes viviparus** (Redfish or Norway Haddock) is smaller than *S. marinus* (1) reaching only 25 cm in length. It is found in Scottish waters usually at depths of 40 – 100 m and is distributed locally. The record rod-caught specimen, caught in the Shetlands, weighed 283 g. The reproductive habits are very similar to those of *S. marinus*. Little is known of its feeding habits, but it eats pelagic fish and crustacea.

3 **Lampris guttatus (luna)** (Opah, Moonfish) is widely distributed in the warmer parts of the Atlantic at depths of 100 – 300 m. Each year a few are observed singly at the surface and others are storm-driven and stranded on the Atlantic coasts of the British Isles. It is a brilliantly coloured, large fish. It may attain a length of 100 – 150 cm; one stranded recently was 109 cm and weighed 44 kg. Its food consists mainly of oceanic squids (their beaks are found in its stomach), as well as crustacea and pelagic fish. The mouth, like that of the John Dory (4), projects forward to take the prey.

4 **Zeus faber** (John Dory or St. Peter's Fish) is common in the English Channel, around Ireland and the Isle of Man, and is also found in the southern North Sea. Generally found in depths of less than 300 m, it swims in mid-water or near the bottom, and during summer it comes closer inshore.
A length of 57 cm and a weight of 8 kg can be reached but 25 – 45 cm is more usual. The record rod-caught specimen weighed 4·876 kg.

Sexual maturity is reached when the male is about 27 cm long and the female 38 cm. Spawning takes place in early summer and the eggs, 2·5 – 2·8 mm in diameter, are adhesive and demersal. Young fish are 43 mm in October and about 130 mm by the next June. It produces sound but whether this is connected with reproduction is not known. It swims slowly undulating the second dorsal and anal fins, and, because of its excessive slimness, it can approach within striking distance of its prey. The jaws then extend forward with very great rapidity to take small fish including pilchards, herrings and sand eels.

Lophius piscatorius (Angler Fish or Frogfish) is found all around the British Isles. It is a bottom-living fish which inhabits deep water although when small it will live in shallower water. It can reach a length of 198 cm. The British record rod-caught specimen weighed 30·899 kg, and the Irish 32·432 kg.
It can assume both the colour and the pattern of its surroundings. Thus well camouflaged it entices prey by means of the lure, a flap of skin which is dark on one side and white on the other. The lure is erected and then jerked to and fro. When an attracted fish rushes towards it, the lure moves downwards so that the victim follows into the angler's huge open mouth. The jaws snap together and the prey is held by the hinged teeth which lie flat until the fish has entered the mouth and then spring up making escape impossible. Many fish are taken by the angler including the cod and haddock.
Spawning takes place in spring and summer. A million or more eggs are laid, 2·2 mm in diameter, each with an outer adhesive, mucilaginous coat forming a sort of matrix in which the eggs are embedded. This structure forms a floating raft which may be 100 cm wide and 1 m long. The young at hatching are about 4·5 mm long and have a large yolk sac. During the period of absorption of the yolk a large conical prominence arises on top of the head. Later long fin rays develop in the region where the prominence was. The pelvic fin elongates into long streamer-like processes. Immature angler fish, 17 – 50 cm long, come inshore along the north-east coast between May and October.

1 NORWAY HADDOCK

2 REDFISH

3 OPAH

4 JOHN DORY, pursuing prey

5 ANGLER FISH

41

REMORA, TRIGGER-FISH, SUNFISH AND PUFFER FISH

The Remora (1), of the suborder Percoidei (p. 28), is a fish with a flat laminated disc on top of the head that is used for attachment to other animals. The order Tetraodontiformes includes 2 – 4, fish whose jaws are often fused to form a beak. Diverse in body-form, they have armour-like scales, prickles or rough skin on the body, and a reduced dentition.

1 **Remora remora** (Remora) has world-wide distribution in warm seas and is occasionally recovered from sharks caught in British waters. The dorsal fin on top of the head has become modified to form an efficient sucker by means of which it attaches itself to another animal, usually a blue shark — on which as many as three remora have been .found — or to a turtle. This habit enables the remora to travel in ease to fresh feeding areas where, on arrival, it detaches itself and swims actively in pursuit of prey. When sated it again seeks a mobile anchorage.

The remora does not breed in northern waters but sites in mid-Atlantic are known where spawning takes place in June and July. The eggs, 1·4 – 1·52 mm diameter, hatch into small fish about 5 mm long. The sucking disc may be seen in young fish of 18 mm and becomes functional when they are 30 – 40 mm long. Adults may reach a length of 60 cm but those found in British waters do not generally exceed 12 – 13 cm.

2 **Mola mola** (Sunfish) occurs in all seas from Australia to the British Isles, although it rarely appears in the North Sea. It has been seen on the surface of Scottish waters, perhaps basking, and a few are washed ashore each year. A large, strangely shaped fish, it can reach a length of 300 cm and a weight of 997 kg but one of 100 – 150 cm is more usual. A sunfish 196 cm long, 262 cm in height between fin tips and 559 kg was washed ashore in Scotland. The record rod-caught specimen weighed 27·214 kg.

The sunfish appears to swim by means of lateral oscillations of the large dorsal and anal fins. The animal tends to yaw from side to side as it moves forward. It feeds upon small crustacea, eel larvae, and small squids.

Eggs are produced in very large numbers, an estimated 3 million by one fish. They are released in the open sea but little is known of the development of the eggs or the larvae except that the latter do not resemble the adults and were at one time considered a separate species.

3 **Lagocephalus (Tetrodon) lagocephalus** (Puffer Fish) is found off the south-west coasts of the British Isles, where it is carried by currents as it is a poor swimmer. A length of 55 cm may be reached but they are usually about 35 cm long. One specimen caught in Irish waters was 47·1 cm long and weighed 910 g.

The puffer fish has the power to inflate itself suddenly with water until it resembles a prickly football, making it almost impossible for a predator to take. The skin of the belly and sides, up to the level of the pectoral fin, is covered with spines. The spines, the ability to inflate, and the striking colouration would together present a formidable appearance to an enemy.

The mouth, moderate in size, has a pair of broad teeth pointed anteriorly to form a beak. Its food consists of fish, crustacea and oceanic squids, the beaks of which are often found in the stomach of puffer fish.

4 **Balistes carolinensis (capriscus)** (Trigger-fish or File-fish) is occasionally captured off the south and west coasts of the British Isles. It is a poor swimmer and is generally found inhabiting rocky coastal areas. Of those reported recently, usually between August and October, the largest was 47·7 cm long and 1·75 kg while the smallest was 21·5 cm and 190 g in weight. The record rod-caught specimen weighed 908 g and was caught off Jersey. The young are more pelagic than the adults which are thought to drift with currents from areas where they are abundant.

The trigger-fish is so called because of the erectile and release mechanisms of the three dorsal spines. These may be raised — perhaps in the face of an enemy. If a captured trigger-fish has these spines erect, then they can only be released from this position by depressing the small third spine. This mechanism is presumably under neural control in the living animal. The mouth is small while the teeth are large and pointed. Little is known of its feeding habits but it may scavenge.

1 REMORA, attached to underside of Blue Shark
2 SUNFISH
3 PUFFER FISH   4 TRIGGER-FISH

THE PLAICE

This commercially important flatfish is a member of the order Pleuronectiformes (p. 46) and the family Pleuronectidae (p. 46).

Pleuronectes platessa (Plaice) lives in the shallow coastal waters of northern Europe. They are most abundant in the North Sea, where the majority of those marketed in the British Isles are taken, and are also found in the Channel, the Irish Sea, and off the Scottish coast.

The life history of the plaice may be followed in the North Sea where concentrations of eggs are found over a wide area during the spawning season which lasts from January until April off Flamborough Head, with a peak in February and March. The eggs, 1·8 to 1·9 mm in diameter (1A), drift in the plankton and take about 20 days at 6°C to hatch (1B). The newly hatched larvae (1C) are 6 – 7 mm long with a large yolk sac attached which is utilized in about 8 days when they are 8·5 mm long. They then begin to feed on diatoms. Dispersal of the eggs and larvae depends upon water currents. In the Flamborough Head area eggs and larvae tend to remain because of eddies.

At hatching the plaice is quite symmetrical. At about 5 weeks of age metamorphosis begins (1E) and is completed by the 7th or 8th week of age, by which time the plaice has come to lie on its left side. The left eye migrates in about 45 days so that both eyes are on the right side which now becomes the upper surface (1F). The head is now remarkably twisted, and the jaws take up an almost vertical position and their musculature undergoes modifications. The pectoral and pelvic fins on the upper side are larger and more posterior than those on the eyeless side. The swimming posture alters as these changes take place. The vertical plane of the body slopes gradually from right to left, following the eye, and in swimming the plane passing through both eyes remains horizontal.

The little fish (1G, 1H) now move shorewards to the nursery grounds where they live and feed on the bottom. Here they spend their first summer in water less than 5 m deep, after which they move progressively into deeper water as they grow larger. Like other flat fish, the plaice is able to adjust its colouration to resemble its background. The pigment is contained in cells that can be dilated or contracted in response to visual stimuli.

Food in the early larval stages consists mainly of diatoms and the spores of algae, changing in the later stages to copepods. After metamorphosis, when the young plaice come to live on the bottom, they take small polychaete worms and copepods, and later small crustacea, gastropods and lamellibranchs as well. From the third year onwards their diet consists chiefly and by preference of lamellibranchs, but also includes crustacea, echinoderms, polychaetes, coelenterates, and fish such as sand eels.

Feeding is seasonal and lasts from March until October after which it almost ceases. This pattern is followed in their growth, which slows or ceases in the winter months. It is this change in growth that provides a means of assessing the age of the plaice, using its otolith, or ear-stone, although in many fish the scales can be used. Microscopic examination of the otolith reveals a series of alternate concentric lamellae of opaque and more translucent material, which form rings or zones. These can be counted to give the age of the fish.

A plaice about 1 year of age is 13 cm in length and 10 g in weight; at 3 years old, 21 cm and 100 g; after 10·7 years, 45·6 cm and 838 g; and at 20 years is about 60 cm and nearly 2 kg. The British record for a rod-caught plaice, of unknown age, is 3·600 kg. Sexual maturity is reached in the male when 3 years old and 20 – 30 cm long, and in the female when 4 – 5 years and 30 – 40 cm. As the females increase in size, so they produce more eggs per unit length; thus, one of 30 cm may produce some 60,000 eggs, and one of 45 cm over 280,000 eggs. It has been estimated that 60 million females will gather in one region of the North Sea to produce some 5,000 billion eggs.

Spawning takes place in water of 4 – 10°C. It is not known whether it occurs on the bottom or in midwater but in an aquarium it took place 75 cm from the bottom. The female lay diagonally across the back of the male, their vents being close together. Quivering violently, the female rapidly emitted a stream of eggs while the male extruded milt. After about 20 seconds the adults separated and settled on the bottom. The female may take a month or more to complete spawning, releasing eggs at intervals of a few days. She would be unable to retain all the eggs at the same time in their final stage because they absorb a lot of fluid. It is likely, therefore, that the eggs ripen in small batches.

Plaice fishery is an important industry and because of intensive and competitive trawling over-fishing is a danger. To investigate the possibility of maintaining or even increasing the population, experiments in farming the sea were made about 1900. Plaice were transplanted to an area rich in food to see if the yield could be increased. More recently it has been found possible to rear plaice to a marketable size, some 225 g, in 2 years or even 18 months if they are kept in a warm water outflow from a nuclear power station. Wild plaice would take about 4 years to reach this size. Survival rates from egg to maturity can be as much as 20 times better than in the sea.

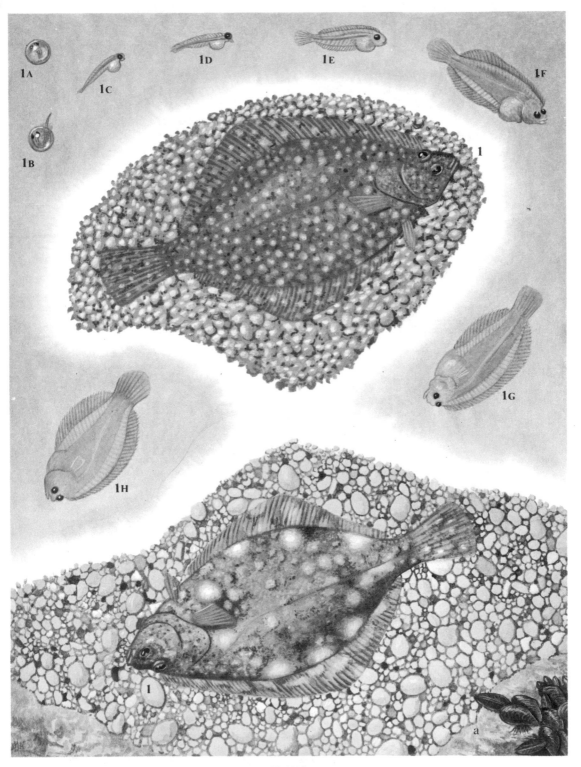

PLAICE

1 PLAICE, showing change in colour to resemble background
1A developing egg   1B egg, hatching   1C newly hatched larva
1D 4 days after hatching 1E beginning of metamorphosis 1F eye has moved over head
1G-1H young bottom-living plaice   a Mussels, an important food of plaice

45

WITCH, HALIBUT AND FLOUNDER

The order Pleuronectiformes includes the flat, bottom-living fish whose eggs hatch into pelagic larvae which are symmetric. After metamorphosis, however, the little fish lie on one side, the upper, eyed side, becoming pigmented while the lower, blind side, usually remains pale. During the transition the bones of the skull twist, the head becomes asymmetric, and one eye migrates to what becomes the upper surface. The swimming posture changes gradually from vertical, at hatching, to horizontal when it becomes bottom-living.

The plaice family, Pleuronectidae, comprises relatively large fish which rest on their left side, the left eye migrating to the right side. The lateral line is almost straight and a spine is present in front of the anal fin.

1 **Glyptocephalus (Pleuronectes) cynoglossus** (Witch) occurs to the west of Ireland and northwards around Scotland and into the North Sea, generally where the bottom is of mud or muddy sand, in water down to 400 m. They may reach a length of 47 cm but 25 – 40 cm is more usual; one female 8 – 9 years, 42·4 cm long, and 473 g was recorded. The record rod-caught specimen weighed 533 g.

Sexual maturation occurs in females when about 27 cm long. The spawning season lasts from May until August. One female of 26·8 cm contained 49,000 eggs and another of 42 cm had 599,000 eggs. The eggs are 1·1 – 1·2 mm in diameter and pelagic. They hatch in 8 – 9 days, the larvae being about 4 mm. Metamorphosis begins when they are about 25 mm long. Growth is rapid for the first 5 years and then proceeds more slowly. A female of 10 years may reach 36 – 40 cm but a male only 33 cm. It is a bottom feeder, feeding throughout the year but the intake varies, being greatest in June and July and smallest during December. One half of the volume of its intake consists of polychaete worms and one third of crustacea. It also eats molluscs and other invertebrates.

2 **Hippoglossus hippoglossus** (Halibut) is an Arctic and sub-Arctic fish. Its southern limit is the English Channel, but it is more abundant in Scottish waters. It is a solitary fish that prefers mixed and rough grounds, particularly gullies between rock ledges and banks. There is a vertical distribution, large halibut being found in water of 50 – 150 m, smaller ones inhabiting shallower waters. The halibut is the largest member of its tribe. A very large specimen caught off Massachusetts was 279 cm long and weighed 284 kg when dressed! The British record rod-caught halibut weighed 73·365 kg. The jaws are of similar size, symmetrical and with teeth almost equally developed on the blind and sighted sides—a condition found in other flatfish that leave the bottom and swim actively. Its food consists mainly of haddock and whiting, but other fish as well as crabs and lobsters are eaten. The male reaches sexual maturity when about 100 cm long, females at 132 cm. Spawning takes place in fairly deep water. The season is prolonged, lasting from January until June or July. Ripe eggs are large—3 – 4·2 mm in diameter. Eggs of this size may be found in a female together with many unripe eggs only 1·9 – 2·0 mm in diameter. Ripening may take place in batches as ripe eggs are shed. Some 3·5 million eggs have been found in a fish of 60 kg. The young when 3 cm long are still pelagic and the eyes are on opposite sides of the head; until about 30 cm long they may be found in shallow water feeding upon young flatfish and shrimps. As they grow larger so they go into deeper water.

3 **Platichthys flesus** (Flounder) is common all round the British Isles in creeks and harbours, where the bottom is of mud and sand. They live in shallow water, and travel with the tide, coming into estuaries on the flood tide. Small ones may be seen in rivers above the tide apparently tolerant of fresh water although they must return to the sea to breed.

Males reach sexual maturity when 11·2 cm long and females when 18 cm. Spawning takes place from February to May in depths of 2 – 15 m in the open sea. A large number of eggs are produced; one female of 710 g was estimated to contain 1,638,000 eggs. The pelagic eggs are 0·95 – 1·03 mm in diameter. They hatch in 5 – 11 days into larvae 2·3 – 3 mm long. Metamorphosis is complete when they are only 10 mm long and, now demersal, lie on one side like miniature adults but their colouration is not acquired until they are about 10 cm long. Pigment on the eyed side appears and guanine develops rapidly to give the blind side its marbled-white appearance (3A).

The young feed on copepods and amphipods initially, and later on crabs, shrimps and fish. The adults eat herring, sand eels, gobies, crabs, shrimps, molluscs, and lugworms. They usually feed during the day while at night they may be found swimming at the surface. The spent adults may be found close inshore from April onwards. They are usually thin but their condition soon improves. As winter approaches they withdraw into deeper channels and gradually move out into deeper water in the open sea, in preparation for spawning.

The flounder can reach a length of 51 cm, one of 44 cm being 5 years of age. The record rod-caught specimen weighed 2·593 kg.

1 WITCH
3A FLOUNDER.underside

2 HALIBUT
3 FLOUNDER

47

LEMON SOLE AND DABS

The fish on this page all belong to the order Pleuronectiformes (p. 46), and the family Pleuronectidae (p. 46).

1 **Microstomus kitt** (Lemon Sole, Lemon) is found all around the British Isles, generally on muddy grounds, and is usually caught in water down to 55 m. It is a commercially important fish and many tons are landed here each year. The best fishing grounds are off the north-east coasts of Scotland and England.

The majority of males are mature by 3 years and most females by 5 years. They spawn when they first reach maturity, in March and April. There is usually a pre-spawning migration northwards by the adults so that the planktonic eggs and the larvae as they hatch drift southwards to suitable feeding grounds. A female of 31 cm may produce as many as 150,000 eggs while one of 38 cm can produce 670,000 eggs. The eggs, 1·13 – 1·45 mm in diameter, hatch in 8 days at 9°C. The larva at hatching is quite symmetric but when 15 mm long metamorphosis has already begun, the left eye moving towards the top of the head. By the time it is 25 mm long it resembles the adult and lies on its left side on the bottom. After metamorphosis the jaws, which remain in their former position, now open and close by a lateral movement.

The prey consists mainly of polychaete worms and crustacea, generally hermit crabs, as well as molluscs and coelenterates. In order to search for this notably sedentary prey the lemon sole lifts its head and the front of its body from the bottom. The prominent, mobile eyes search the area and if a worm moves the fish pounces, arching its back to bring the jaws almost vertically onto the prey. Feeding is seasonal; in December the lemon sole stops eating and resumes again in February. The peak period of food intake is May or June after which there is a gradual decline. The best rates of growth are achieved when polychaetes, crustacea, lamellibranchs, gastropods and coelenterates are all available. Lemon soles of up to 48 cm occur regularly in trawls. The largest one recorded measured 67 cm and was captured in 1907. A female 23 years of age and 57 cm long weighed 2·949 kg when gutted. A male of the same length weighed 2·230 kg but was only 15 years of age. The record rod-caught one weighed 990 g.

2 **Hippoglossoides platessoides (limandoides)** (Long Rough Dab) may be found southwards from the Arctic to the English Channel, being common to the west and north of Scotland. It also occurs in the western Atlantic. It prefers a bottom of mud and is usually caught at depths of 30 – 60 m.

Generally 15 – 25 cm in length, one of 30 cm weighs about 190 g, a fish of this size being 5 years old or more. They are probably sexually mature when about 10 cm long. Spawning fish have been found at depths of 12 – 50 m; the main period lasts from February to May with a peak in March. The eggs are pelagic and the capsule quite large, 1·8 mm in diameter, but much of this is taken up by space surrounding the yolk, the egg itself being about 1 mm. Hatching occurs in 5 – 14 days and the larvae, about 4 mm long, apparently go to the sea bottom in deep water shortly afterwards. Larvae 5 – 13·5 mm long have been caught from March until May off the west coast of Ireland.

3 **Limanda limanda** (Dab) is found from Iceland southwards to the Bay of Biscay and the southern North Sea. Dabs prefer a bottom of mud or sandy mud and live in water to a depth of 100 m. Small dabs are often found close inshore and somewhat larger ones will enter estuaries.

The mouth is small, the cleft extending further back on the blind side. The teeth are weaker than those of the flounder or plaice but the eyes are larger. These are used in finding prey as the dab feeds during the day and attacks when the prey moves. It eats crustacea, particularly shrimps, lamellibranchs such as the razor shell *Ensis*, and echinoderms. Some food is probably located by smell.

Although much of its life is spent upon the bottom, the dab can swim quite rapidly for short distances and one 7 cm long has a speed of 9 cm per second. Sexual maturity is reached at about 2 years of age when they are about 15 cm long. Spawning takes place at depths of 20 – 40 m from February to July, with a peak from March to May. One female 22 cm long can produce some 128,812 eggs. The spent adults move inshore to feed and recover after spawning. The eggs, 0·8 mm in diameter, are pelagic and hatch in about 12 days at 12°C. The larvae are 2·7 mm long at hatching and ones of 2·75 – 4 mm have been caught in February and March off the west coast of Ireland. Shortly after hatching they tend to go to the bottom. Metamorphosis begins when the larvae are about 7 mm long and is complete when they are 18 mm. Adults reach a length of 22 – 30 cm when they are about 5 years of age. The record rod-caught specimen weighed 1·211 kg.

1 LEMON SOLE
2 LONG ROUGH DAB
3 DAB

SOLE AND FAMILY

The fish on this page belong to the order Pleuronectiformes (p. 46) and to the family Soleidae. Their eyes are on the right side and the rounded snout projects beyond the downwardly curved mouth. Of all the flatfish these are the most highly specialized and completely adapted for life on the bottom.

1 Solea solea (Sole) is the commonest member of the family in British waters, found in the Bristol Channel, English Channel, Irish Sea and the North Sea south of the Firth of Forth. A fish of the continental shelf. it lives mainly in shallow waters particularly in the vicinity of rock ledges and reefs. Sole may be caught at depths of 4 – 120 m but most frequently at 10 – 80 m, where the bottom is of fine sand, sandy mud, or mud. The sole, like other flatfish, can adjust the chromatophores on the upper surface so that its colour resembles the background.

Sexual maturity is reached in males when about 20 cm and in females when 26 cm, but the size may vary from area to area. During the period that precedes spawning soles have, on occasion, been seen swimming at the surface in quite large numbers, but the reason for this behaviour is not known. The spawning season in the Plymouth region is from February to June and in the North Sea from April to August, usually at depths of 46 – 55 m. A sole of 500 g may produce some 134,000 eggs. These are pelagic, 0·9 – 1·5 mm in diameter, with many oil globules arranged in groups. Hatching occurs in about 10 days and larvae 3·2 – 3·7 mm long emerge. They are pelagic and drift into shallower water feeding on copepods such as *Temora* and also decapods and fish larvae. Metamorphosis begins when the larva is about 7 mm long and is completed when it is 12 mm. Post larvae of 4·5 – 5 mm have been caught in March and August off the west of Ireland, and their planktonic life seems to last quite a long time. By the end of a year the young fish is 15 cm long. Adults are usually 30 – 45 cm in length. The record weight for a rod-caught specimen is 1·866 kg.

Feeding chiefly at night, the sole depends almost entirely on its sense of touch. It creeps along the bottom with the aid of the fringe of fin rays, meanwhile moving its head upwards and sideways and patting the ground gently at intervals. A dense mass of papillae on the blind side of the head feels objects in the substratum and food, when located, is immediately seized. The prey consists chiefly of polychaetes and crustacea, but fish, molluscs and echinoderms are also taken. The sole generally remains on the bottom partially buried in sand during the day, but it may occasionally respond to prey visually although the eyes are very small and almost immobile.

2 Pegusa lascaris (Sand or French Sole) is found from the Mediterranean to the south-west of the British Isles and it just strays into the North and Irish Seas. An inhabitant of inshore shallow waters, 3 – 18 m, deep, where the bottom is of sand, it may be common in some areas. In July and August some, 11 – 18 cm long, have been found to enter rivers and feed upon crustacea.

Spawning takes place from May to August with a peak in July. The eggs, about 1·37 mm in diameter, have numerous oil globules arranged in groups. The larvae, 4 mm long are still quite symmetric at 8 mm but metamorphosis is complete when a length of 22·5 mm is reached. The larvae feed mainly upon small crustacea. The teeth are well developed on the blind side. Adults can attain a length of 35 cm.

3 Microchirus variegatus (Variegated or Thickback Sole) is found from the Mediterranean to the south-west of the British Isles where it has a restricted range. It occurs off the Isle of Man, the Eddystone grounds near Plymouth, and along the west coast, occasionally straying into the North Sea. It inhabits water 40 – 320 m deep, where the bottom is of sand, mud, or shell gravel. Spawning takes place from February to July at Plymouth and from June to August around the Isle of Man. The eggs, 1·28 – 1·36 mm, have numerous oil globules and at hatching the larvae are 2·4 mm long. Ones of 4 – 6 mm have been caught off the west coast of Ireland between May and September, and off Plymouth from March to August. These young fish feed mainly on copepods until they are about 8 mm long. The left eye begins to move forward and upwards when the young fish are about 9 mm long and metamorphosis is complete when they are 43 mm. The adults do not exceed a length of 21·5 cm, the female being slightly larger than the male.

4 Buglossidium luteum (Microchirus boscanion) (Solenette) has a distribution extending from the Mediterranean northwards to the English Channel and the southern North Sea where it is common. It has been caught in water 6 – 76 m deep where the bottom is of sand.

The male is sexually mature at about 7 cm and the female at 9 cm, when they are probably about 2 years of age. Spawning occurs from March to July. The eggs, 0·7 – 0·9 mm, are pelagic and have many oil globules. They hatch in 5 – 6 days and the larvae, 1·8 – 2·3 mm long, remain in the plankton until they are 9 – 10 mm when metamorphosis is complete and they go to the bottom. By the end of their first year the fish are 4·6 cm. They can attain a length of 12·5 cm.

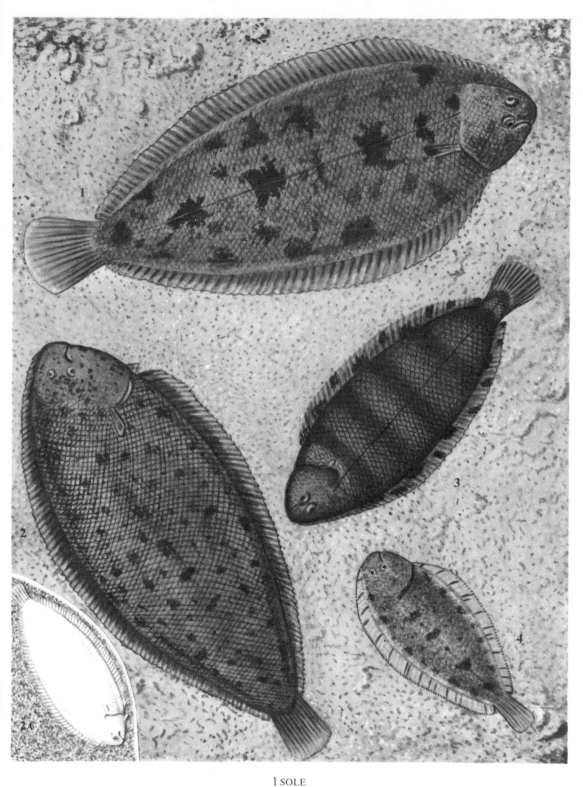

1 SOLE

2 SAND SOLE
2 A SAND SOLE, underside

3 VARIEGATED SOLE
4 SOLENETTE

TOPKNOTS AND SCALDFISH

The fish on this page belong to the order Pleuronectiformes (p. 46) and the family Bothidae. Flatfish of this family all have their eyes on the left side and lie on their right side. None is present in the abundance of other flatfish and only the turbot, brill and megrim (p. 54) are relatively common and commercially important.

1 **Phrynorhombus norvegicus** (Norwegian Topknot) is found in the southern North Sea, the Channel, in Irish waters and off the Isle of Man. It generally lives at depths of 10 – 170 m on rough ground, in areas with underwater cliffs where it holds itself vertically on the rock face. It feeds mainly on crustacea.
Sexual maturity is reached when about 7 cm long. Spawning occurs from March to June off the west coast of Ireland and Plymouth. The eggs are pelagic, 0·72 – 0·98 mm in diameter, and they hatch in about 6 days. The larvae and post larvae have been caught in Irish waters from May to July when they are 4 – 6·5 mm long; at Plymouth they occur from March until August. The post larvae make diurnal migrations being between 5 and 25 m at dawn but remaining below 10 m at dusk. Metamorphosis begins when they are about 9 mm long and is complete by the time they are 13 mm long. The adults may attain a length of 11·5 cm.

2 **Arnoglossus laterna** (Scaldfish) is present all around the British Isles but in Scottish waters it is very rare, although it has been recorded there in recent years. It is an inhabitant of inshore waters with a preference for areas where the bottom is of sand and the water 10 – 100 m in depth.
Sexual maturity is reached in the male when about 11 cm long, and later, when 13·2 – 14·7 cm in length, there is a development of the second to sixth dorsal fin rays. The females mature sexually when about 14 cm in length and there is an incipient development of the male characters. Spawning occurs from April to June to the west of the British Isles and from May until August in the North Sea at depths of 10 – 40 m. The eggs are pelagic, 0·6 – 0·76 mm in diameter, with a single oil globule. At hatching the larva is 2·6 mm long. Post larvae 3 – 6 mm long have been caught from June to September off the west coast of Ireland. The pelagic larval life may last for a long time; it is thought that the first winter is spent at this stage. Metamorphosis takes place when they are 16 – 30 mm long. The adults can reach a length of 19 cm.

3 **Zeugopterus punctatus** (Common Topknot) has a range from the Bay of Biscay to the west coast of Norway. It is found all around the British Isles but with most frequency off the south-west coasts in shallow water, about 37 m deep, on rocky grounds. It sometimes clings to rock surfaces by flexion of the body and it is well concealed by its colouration. Spawning occurs off Plymouth from February to March and a little later off the west coast of Ireland. The egg, 1 mm in diameter, has a single oil globule. The larva is 2·5 mm in length and has yellow dots on the fin. The post larvae have been caught in depths of 20 – 60 m between April and June when 3 – 8 mm in length. By January they have reached a length of 4 cm, have adult colouration, and live on the bottom. The adults can reach a length of 25 cm. The record weight for a rod-caught specimen is 228 g.

4 **Phrynorhombus regius** (Eckström's Topknot) is only rarely found, and appears to be confined to the west coasts where it has been found in offshore waters, sandy bays, over rough ground and in estuaries.
It spawns from April to June off Plymouth and from March to June off the west coast of Ireland. The eggs, 0·96 mm in diameter, are pelagic. The incubation period lasts for 6 days. The post larvae have been caught from May to July in deep water layers off the west of Ireland.

Arnoglossus imperialis has been recorded in Irish waters, the south-west coast of which represents the northern limit of the species. It has also been taken off Plymouth. Pelagic young have been found off the Scottish coasts at depths of 24 – 100 m between September and November although no adults have been found there. The adults may reach a length of about 20 cm.

Arnoglossus thori is occasionally trawled from rough grounds near Eddystone in the Channel. The adults can reach a length of 15 cm while its eggs are 0·72 – 0·74 mm in diameter.

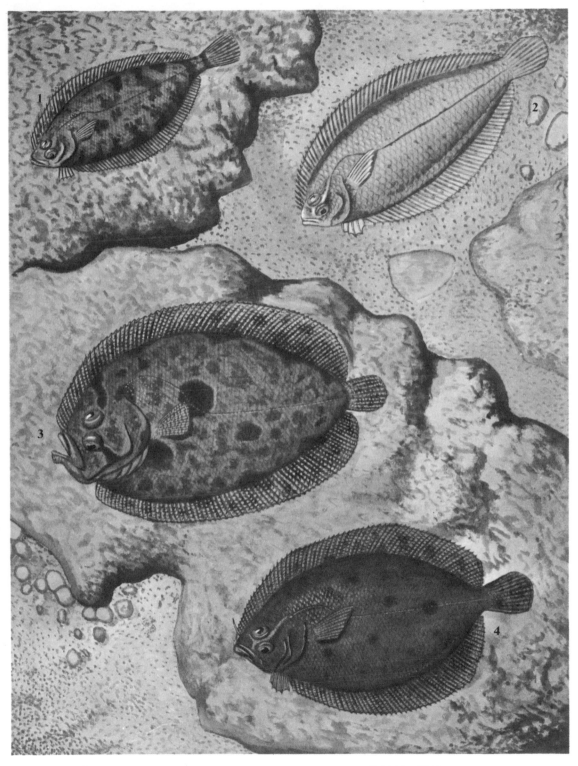

1 NORWEGIAN TOPKNOT   2 SCALD FISH
3 COMMON TOPKNOT   4 ECKSTRÖM'S TOPKNOT

THE TURBOT AND FAMILY

The fish on this page belong to the order Pleuronectiformes (p. 46), and the family Bothidae (p. 52). The turbot is one of the most valuable and highly regarded of the flatfish.

1 **Scophthalmus rhombus** (Brill) is common in the south-west North Sea, the Channel, and off the south and west of Ireland and the west of Scotland. As it has the same geographical range and likes the same type of ground as the turbot (3), they are often caught together. The majority are found in depths of about 40 m and their limit is 100 m. The brills' seasonal migrations are also similar to those of the turbot. The adults are usually 5 – 6 kg but the record rod-caught specimen weighed 7·257 kg.

Spawning takes place from March to August in the North Sea. The pelagic egg, 1·2 – 1·5 mm, has a single oil globule. A female of 2·5 kg produces about 825,000 eggs. Newly hatched larvae, 4 mm long, are pelagic until about 28 mm in length. The stages of metamorphosis are similar to those of the turbot except that brill larvae are larger at all stages. Young brill, 7 – 26 cm long, captured in estuaries, were found to be feeding largely upon the sand goby (p. 82), together with other small teleosts, and also *Crangon vulgaris* and other crustacea.

Hybrids of the brill and turbot are known, both from the sea and after artificial fertilization in the laboratory. The offspring tend to be intermediate having characteristics of both fish.

2 **Lepidorhombus whiff-iagonis** (Megrim) is a fish of the north-east Atlantic from the Mediterranean to the North Sea. It is common in Irish and Manx waters, west and north of Scotland, and in the northern North Sea. Essentially a deep water species it is usually caught in depths of more than 110 m, generally over a soft substratum, but it can be found between 8 and 440 m. These flatfish migrate from their winter headquarters off the north of Scotland into deep water in the North Sea, returning in the autumn.

The males are probably sexually mature at about 25 cm long and the females at 30 cm. Spawning occurs between March and June at depths of 100 – 400 m. The pelagic eggs, 1·1 – 1·5 mm in diameter, have a single oil globule, and hatching takes place in 6 – 7 days. Metamorphosis is complete when a length of 25 mm is reached and the young fish are then benthic. A length of 59 cm can be attained but they are generally between 28 and 37 cm. The record rod-caught specimen weighed 1·644 kg.

3 **Scophthalmus maximus** (Turbot) is found in the southern North Sea, the Channel, off the south and west coasts of Ireland, and occasionally off the west coast of Scotland. They are valuable commercial fish and the majority are caught in depths of less than 110 m but some may go down to 160 m. They appear to migrate seasonally, inshore in summer and offshore in winter. From tagging experiments it has been found that immature turbots make long migrations into deeper water. They prefer a bottom of sand, mud, or gravel, and favourite haunts are banks and estuaries where sand eels and sprats are present as these fish form a large proportion of their food. Small flatfish, whiting, dragonets and herring are also eaten.

The minimum size at which sexual maturity is reached is 30 cm in males, and 35 cm in females, but in some areas the fish are larger. Spawning occurs off-shore in water of 10 – 40 m depth from April to July. The pelagic eggs, 0·9 – 1·2 mm diameter, have a single oil globule. It has been estimated that a female of 6·4 kg may produce about 8·5 million eggs and one of 10·4 kg, 14·3 million. Hatching takes place in 7 – 8 days and the larva, 2·2 – 2·8 mm long, has a very large yolk sac which is utilized by the time it is 3·2 – 4·8 mm long. Once the yolk has been absorbed the little fish swims with rapid strokes of the tail and vibration of the pectoral fins, darting at small crustacea if they move. Metamorphosis begins when the larva is about 5 mm and is complete when it is 20 mm. A swim bladder develops during the pelagic stage but is lost when the fish is 50 mm long and has become entirely demersal. At this size it lives in very shallow water sometimes even between the tide marks. There is a gradual movement into deeper water as it grows larger.

Post larvae that hatched in spring or early summer are caught at the surface when about 6 mm long and are still pelagic or partly so when 19 – 25 mm in July. By August they are bottom-living and a month later are 41 – 62 mm long. In December they are about 7 cm long and by the following June, when about a year, are 10 – 15 cm.

Adult males may attain a length of 53 cm and females 68 cm at 15 years of age. The majority weigh about 5 kg but the record rod-caught specimen weighed 13·153 kg.

Experiments in rearing turbot from egg to marketable size have been made and although the adults spawn well in captivity it has proved difficult to keep and feed the young.

1 BRILL   2 MEGRIM

3 TURBOT

GOBIES

These fish, members of the order Perciformes (p. 24) and the suborder Gobioidei, are generally small. Often found together in large numbers, they inhabit the sea-bed in offshore waters — except for the transparent gobies (2 and *Aphia minuta*) which are pelagic. Gobies lay relatively large, adhesive, club-shaped eggs generally under stones, in empty worm tubes or mollusc shells, at the bottom of the sea. In some species the male guards the nest until the eggs hatch. The young are planktonic. The distinctive feature of these fish is the fusion of the pelvic fins by a membrane which extends across the fins anteriorly. This structure forms a mechanism for attachment to the substratum (see 4), although not a strong one when compared with those of the fish on pp. 78 and 80.

1 **Lebetus orca** (Diminutive Goby) has been found in Manx and Irish waters and occasionally off Plymouth. It lives on coarse grounds at depths of 2 – 375 m. Sexual maturity is reached at 1 – 2 years. In the male it is accompanied by a change in colouration and in length of fin rays. His body is yellowish or dusky grey with grey bars, while the head is reddish grey above and suffused with reddish orange below, as is the breast. The first dorsal fin is dusky yellow edged with white and the second has an intense black edge. All the females and immature males have the same markings, but the colour is less dramatic than in the mature males. Breeding takes place from March to August. Between 140 and 270 ripening oocytes have been found in one female. Young stages have been caught throughout the summer. The maximum length attained is 4 cm and their life-span is of at least 2 years. These fish are predators and feed upon demersal animals, mainly crustaceans and polychaetes.

2 **Crystallogobius linearis (nilssoni)** (Crystal Goby) is found all around the British Isles though less often off the south-eastern coasts. It is pelagic, living mainly offshore in depths of between 20 – 80 m, but can be found at 10 – 400 m. This is one of the two almost transparent gobies the other being *A phia minuta* (see below).
The male has both more pigment, and a deeper and more laterally compressed body than the female, which has a slender body with pointed head and toothless jaws. The first dorsal fin, which has two fin rays, is present only in the male. The male reaches a length of 5 cm and the female 4 cm. They probably live for a year and die after spawning, which occurs between May and August. One female may produce some 3000 ovoid eggs, each 1·2 – 1·5 mm long and 0·5 mm wide. They are laid, usually at a depth of 30 m, in the empty tubes made by the polychaete worm *Chaetopterus*, and the male guards them. The larvae are in the plankton from May until August. By September they are 2·2 cm and one or two months later 3 cm long. They feed on copepods and other small organisms.

3 **Buenia jeffreysii** (Jeffrey's Goby) is found in the Firth of Clyde area, in Manx waters, off the south and south-west coasts of Ireland and off Plymouth where it is common. It is an off-shore deep-water species inhabiting depths of 38 – 360 m.
Spawning takes place off Plymouth in July and August, the eggs being deposited in empty shells in water 40 m or more in depth. Young of 4 – 9 mm have been captured in July, and by September ones of 5 – 32 mm are found; the larger ones now go to live on the bottom. Adults reach a length of 5 cm.

4 **Lesueurigobius friesii** (Fries's Goby) occurs frequently in Manx waters but is scarce in Irish waters. It inhabits deep water offshore, 10 – 120 m, where the bottom is of soft mud, muddy sand or gravel. Recently it has been reported that it lives in association with the Norway lobster or Dublin Bay prawn which burrow into the substratum. It seems likely that the goby enters the burrow to escape predators and perhaps also fishing tackle.

Gobius fagei can be found in Manx waters at depths of about 95 m, usually on muddy sand. A single post larva has been caught in June.

Aphia (Aphya) minuta (White Goby) like the Crystal Goby (2) is transparent but in contrast is found in inshore waters usually at depths of 3 - 50 m. These fish live in shoals and generally swim near the surface. They have small, fragile scales, and the dorsal fin has five well-developed rays. A length of 6 cm can be attained.

Gobius cobitis (Giant Goby) has on rare occasions been found off the south-west coasts, mainly along the Cornish coast and off the Channel Islands. Southwards it extends into the Mediterranean. It is an inhabitant of shallow pools in the upper half of the shore, where there are rocks or boulders and gravel on the bottom, and green algae. The algae forms their chief item of food, but they also eat crustaceans and other small animals. Spawning occurs in May and June in Brittany. The females produce many eggs; one of 16 cm may have 8,000 – 12,000 eggs, each 1·0 – 1·2 mm in diameter. In large rock pools near Plymouth one often finds individuals 23 cm long, and these are about 9 years old. A length of 25 cm can be reached.

1 DIMINUTIVE GOBY, Female 1A DIMINUTIVE GOBY, Male   2 CRYSTAL GOBY
3 JEFFREY'S GOBY
4 FRIES'S GOBY

PERCH-LIKE FISH

The fish on this page all belong to the order Perciformes (p. 24) and all except 2 and 3 belong to the suborder Percoidei (p. 28). The weever fish (2, 3) belong to the suborder Trachinoidei.

1 **Dicentrarchus (Morone) labrax** (Bass), a warm-water species, is common along the west and south coasts of Ireland and the south-west and Channel coasts of England, northwards in the Irish Sea to the Mersey, and in the North Sea south to Lincolnshire. The adults are usually found inshore on rocky coasts and in summer over ground of rocks, sand and mud in brackish waters and estuaries.

Males reach sexual maturity when 5 – 6 years of age, about 34 cm long and 540 g in weight; females probably mature later. The breeding season lasts March to mid-June in Irish waters. Eggs are pelagic and can be taken in surface waters, close to shore or in the mouths of estuaries. Spawning thus takes place, at least sometimes, in inshore waters with swift tides. The spawning period is prolonged and eggs are likely to be found where shoals of bass are feeding. The eggs are probably shed in batches as they ripen. They are 1·2 – 1·4 mm in diameter and have one or more oil globules which fuse together during development. In an aquarium the eggs hatch in 4 days. Just before hatching the larva wriggles until it bursts the capsule and emerges tail first, 3·8 mm long, almost transparent, and still with a large yolk sac. In 6 days, when 7 mm long, it begins to swim and pigmentation develops. After the yolk has been utilized, the post larvae feed upon smaller planktonic animals. Bass grow slowly, mainly during summer. They have a long life — a female of 76 cm and 5·326 kg was found, by examination of the otoliths, to be 21 years of age. Most fair-sized and large bass are females, although males of up to 2·750 kg and 61 cm have been recorded. Adult bass shoal when feeding, preying on crustacea, particularly *Crangon vulgaris* and mysids, but also on other animals such as small teleosts and polychaetes. They compete with the flounder (p. 46) for food. The record rod-caught bass weighed 8·220 kg.

2 **Trachinus draco** (Greater Weever) less common than *T. vipera* (3), inhabits rather deeper water although it sometimes enters inshore waters. It has been found in Irish waters, in the English Channel, once in Manx waters, and recently in Scottish waters where the females caught from March till May were maturing, ripe, or spent. Spawning occurs in Plymouth waters from June to August. The eggs, 0·94 – 1·11 mm, have a single oil globule and hatch in 7 – 8 days. The newly hatched larvae are about 3·3 mm long.

Adults may be 45 – 50 cm and one caught recently in Irish waters was 39 cm long and weighed 377 g. The record rod-caught one was 1·02 kg. Some fishermen consider this fish a delicacy.

The colouration is perhaps a warning one for this fish can injure the unwary bather who steps on it on sandy ground. The poison apparatus consists of spongy glandular tissue at the base of the 2 opercular spines and 5 – 8 dorsal spines. Each spine is covered by an integumentary sheath, the rupture of which allows toxin to escape from the venom-containing cells along the grooves on the spines. The pain from a sting is intense, increasing in severity for 20 – 50 minutes and lasting 16 – 24 hours if medical treatment is not obtained.

3 **Trachinus vipera** (Lesser Weever) is probably present all around the British Isles and is common off Plymouth, in Manx waters, and in the North Sea. These coastal fish inhabit shallow water with a sandy bottom and even sandy areas far from the coast such as the southern North Sea. There is a migration inshore for summer and offshore in winter to deeper water. The weever is well adapted for lying in wait, buried in sand, as its eyes and mouth are on top of the head. Its prey is chiefly crustacea, small fish such as the sand goby (p. 82), sand eels (p. 60), and flatfish, as well as lamellibranchs and polychaetes. Like *T. draco* it has a poison apparatus and can inflict painful stings.

Spawning occurs in moderate depths and the eggs, 1·1 – 1·37 mm, have a large number of yellow oil globules. They hatch in about 10 days. The larvae, 3·3 mm long, remain in the plankton until 15 mm when the adult characteristics begin to appear. At 30 mm they become benthic, burying themselves in the sand. Adults rarely exceed a length of about 11 cm.

Epinephelus guaza (Dusky Perch) is only rarely recorded. Small ones have been taken in trawls. One specimen recently reported as the first to be taken in Irish waters was 93 cm long and weighed 10·489 kg when gutted. It is now in the National Museum, Dublin.

Polyprion americanus (Stone Basse, Wreckfish) is occasionally found in inshore waters, mainly off the south and west coasts of the British Isles. One immature male was, however, taken in Scottish waters; it was 54 cm long and weighed almost 6 kg. Another, found inside a floating tea chest, was kept alive in an aquarium more than 6 years, and grew from 50 to 60 cm in length. In the sea it has a habit of accompanying floating wood or seaweed. A large fish, it can attain a length of 200 cm and a weight of 45 kg or more.

Serranus cabrilla (Comber) occasionally reaches the south-west coast of England where it may be taken in lobster pots. The record rod-caught one was 294 g. It spawns in July and August and post larvae are found from August to October. Adults may be 25 cm long.

1 BASS
2 GREATER WEEVER
3 LESSER WEEVER

SAND EELS

These fish belong to the order Perciformes (p. 24) and are members of the suborder Ammodytoidei. They are elongate, eel-like fish that inhabit areas where the bottom is of sand in which they spend much of their time in burrows. All are small, the largest (1) reaching a maximum length of 38 cm. Scales when present are small and only on the body. They are an extremely important part of the food chain of the sea as so many fish and sea birds prey upon them. Man also catches them in large numbers to make fish meal. They are amongst the few marine fish that lay their eggs on the sea-bed like the herring (p. 20).

1   **Hyperoplus (Ammodytes) lanceolatus** (Greater Sand Eel) is common around the British Isles. Shoals may be found in sandy areas at depths of 24 – 40 m.
Sexual maturity is reached in 2 years. There is a prolonged spawning period and ripe and running fish are found from April to August. The number of ova increases as the female grows larger; ones of 24·2 cm and 30·6 cm produce 37,527 and 62,068 eggs respectively, each egg being about 0·76 mm in diameter. Larvae are found from April to September, with a peak in June in the southern North Sea. Ones up to 36 mm long are caught in September.
Juveniles feed on zooplankton but the adults are fish eaters, taking the eggs, larvae and adults, often of other species of sand eels. They may reach a length of 38 cm, a weight of 130 g and live for 9 – 10 years.

2   **Ammodytes tobianus** (Lesser Sand Eel) is common all around the British Isles, usually in inshore waters where the bottom is of sand. Spawning takes place from August to October off Plymouth, and in the latter part of the year in the northern North Sea, at depths of 20 – 80 m. The eggs are somewhat irregular in shape, 0·8 mm long and 0·7 mm wide, and have a single oil globule. Females 14 cm and 18 cm long have been found with 13,350 and 36,305 eggs respectively. The egg capsule is thick and adheres to sand grains, although if currents dislodge them they may become planktonic. Larvae may be found in the latter part of the year but they are present in large numbers at the beginning of January. At hatching the young fish are 4 –5 mm long and inhabit the bottom. Larvae 10 mm in length move gradually towards the surface waters which they reach when about 15 mm long. The mid-water existence of the adults is adopted by the young fish when 30 mm long. During the summer there is an inshore migration but they depart to deeper water for the winter. They feed upon copepods, fish and annelids.

3   **Ammodytes marinus** (Sand Eel) is common around the British Isles, in Manx, Irish, and Scottish waters and occasionally in the western part of the English Channel. It is a shoaling fish found concentrated in well-defined areas, at depths of 24 – 40 m, where the bottom is of sand the particles being of such a size as not to hinder ventilation of the gills.

The majority reach sexual maturity at 2 years, a few at 1 year. Spawning takes place from December to February and at this time more females than males are present in a shoal. The eggs are laid on the bottom and the larvae appear suddenly and in very large numbers which suggests synchronous spawning. They are found in the plankton in the Straits of Dover at the end of January and in the north-western North Sea in March. The number of larvae of this species may indeed at times almost equal the total of all young fish in the plankton. The sand eel, when it first begins to feed, takes phytoplankton but as it grows the diet becomes varied and includes appendicularians, larval invertebrates and copepods, chiefly *Temora* and *Calanus*. The adults feed mainly on annelids, newly hatched polychaetes, and crustacean larvae. There is a marked diurnal rhythm. At dawn they begin to feed intensively for 2 – 3 hours, and then at a diminished rate until darkness. The sand eel may then bury itself in the sand, or swim in midwater or even at the surface.
Those found off the Dogger Bank are larger than those taken further south. A length of 25 cm and a weight of 32 g may be attained in a fish of 4 – 5 years. The oldest specimen recorded was 9 – 10 years old. This species forms 95 per cent of the catch of sand eels used in the production of fish meal. They are caught during the day when feeding.

4   **Hyperoplus (Ammodytes) immaculatus** is found in Manx, Scottish and Irish waters, the western English Channel and the North Sea. The time of sexual maturation is not known but it spawns from January until April. The juveniles and adults are found in estuaries, on shell gravel and over rough ground.

5   **Gymnammodytes semisquamatus** (Smooth Sand Eel) occurs off most of the coasts of the British Isles. It is a shoaling fish which inhabits sandy areas at depths of 30 – 40 m. The majority are sexually mature by the end of one year. The spawning season is long — from March until August with a peak in June and July. The larvae are found from June to November in the southern North Sea, but this and the spawning period vary with the area, being later in the northern North Sea. In June most larvae are 3 mm long, by July 5 mm, and in September 17 mm. The adults reach a length of 23·5 cm and a weight of 28 g. They live for 7 – 8 years.

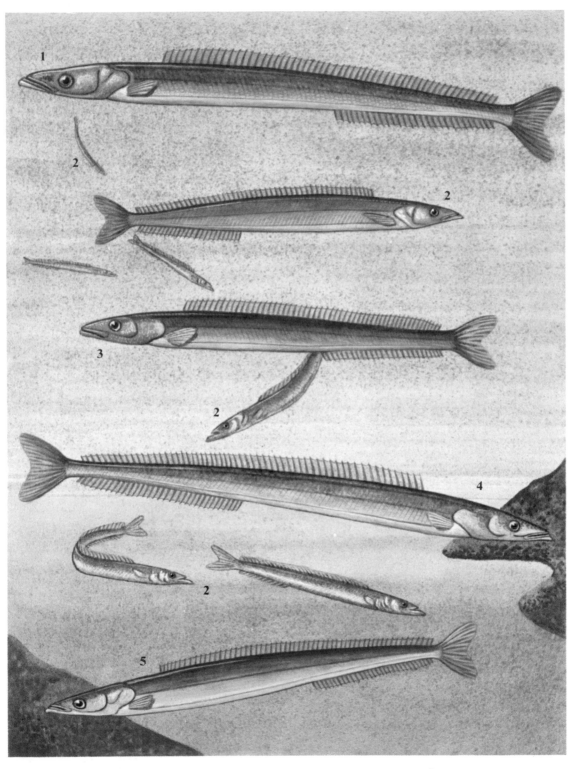

1 GREATER SAND EEL

3 SAND EEL

2 LESSER SAND EEL

4 HYPEROPLUS IMMACULATUS

5 SMOOTH SAND EEL

WRASSE

Fish on this page belong to the order Perciformes (p. 24), and the suborder Labroidei. They have thick lips and strong conical teeth. The gill covers and cheeks are covered with hard scales. Brightly coloured fish, they inhabit shallow water near to rocks, some even being intertidal. They display sexual dimorphism. The males of some species build a nest and guard the adhesive eggs, others have pelagic eggs. They move through the water by synchronous rowing strokes of the pectoral fin while the tail fin is used as a stabilizer and rudder.

1 Centrolabrus exoletus (Rock Cook) occurs occasionally in British waters but is a northern species better known in Scandinavian waters. It is small, not exceeding a length of 15 cm.

2 Coris julis (Rainbow Wrasse), common in the Mediterranean and adjacent Atlantic, appears only rarely in British waters. It has no scales on the head or gill cover. A length of 32 – 35 cm may be attained. The small egg, 0·65 mm, is pelagic, and the newly-hatched larva is 2·5 mm long.

3 Labrus mixtus (ossifagus) (Cuckoo Wrasse), common in Irish and Manx waters, sometimes occurs off Plymouth and in Scottish waters. It lives near rocks usually at depths of 40 – 180 m. A male observed in an aquarium built a nest, beginning in May, and then would dart at females until one showed interest when he spread all his fins fully to show the vivid colouration. With mouth open, head and shoulders were jerked from side to side, while his body was twisted into an S-shape. This display induces the female to deposit eggs in the nest. Males are usually under 35 cm long, females (3A) 25 cm. The record rod-caught specimen weighed 680 g.

4 Labrus bergylta (Ballan Wrasse) can be found in the Mediterranean and adjacent Atlantic from north-west Africa to Norway. It is fairly common, particularly from May to November, in British coastal waters of shallow or moderate depths where there is rock and a weed covered bottom. Although not a shoaling fish, many appear together on occasion. Larger fish are found at depths of 10 – 20 m, small ones close inshore in shallow water. During winter they all withdraw to deeper water but in very severe winters they may appear, semi-torpid, inshore.
Diurnal in habit, they sleep sprawled on their sides or leaning against rocks. Females mature sexually when about 25 cm long and 4 years old but some at 5 years. Spawning occurs in June and July, possibly even in May. Nests, wedged in rocks, are formed from tufty seaweeds bound together with mucus. The eggs, 1·1 mm in diameter, adhere slightly to the seaweed. At hatching the larvae are 3·8 mm long and wriggle actively. They remain in the plankton during summer and measure some 28 mm by September and 45 mm by October. They grow slowly; at 2 years they are 18·4 cm and at 10 years 40 cm long. When 30 – 45 cm they weigh 3 – 3·5 kg. The record rod-caught specimen was 3·485 kg. Their food consists mainly of crustacea including crabs and prawns, but also lamellibranchs, ragworms, and *Spirorbis*.

5 Crenilabrus melops (Corkwing Wrasse) is common in Irish and Manx waters, is found in the English Channel and extends into the southern North Sea, but is rare in Scottish waters. It likes rocky, weedy areas in shallow water particularly in summer, and may be found in harbours.
The female matures when about 3 years of age and 15·3 cm long. In spring the mature males become brightly coloured and display territorial and nest-building activities. When the nest, formed of three layers of algae, is complete, the male shows fanning activity before courtship and spawning. One male may attend 4 to 13 nests all at different stages of completion. Many nests are in the intertidal zone and the fish have to move in and out with the tide. There is a change in territorial aggression during such enforced absence from the nest. Aggressive behaviour is exhibited by the male during the reproductive period. This is modified when a ripe female approaches the nest and lays eggs at the entrance. The eggs, about 1·8 mm in diameter, are inserted into the nest by the male, who rubs himself against it as he fertilizes the eggs. After many repetitions of this spawning activity, the female is chased from the nest. The male guards and fans the eggs for 11 – 16 days until they hatch. He begins building a nest again about a week later, ready for the next spawning cycle.
Young corkwings have been caught in September when 17 – 25 mm long, others of 35 – 43 mm were captured in October and by the following March ones of 55 – 77 mm have been found. Typical adult colouration has developed by the time a length of 40 mm has been reached. The adults are usually 15 – 20 cm long but they can reach 30 cm, indeed ones of 28·5 cm have been caught recently in Scottish waters. The record rod-caught specimen weighed 187 g. They feed upon *Spirorbis*, ragworms, barnacles, gastropods and amphipods.

6 Ctenolabris rupestris (Goldsinny) has a range from the Mediterranean to the Baltic and has been recorded from various parts of the British Isles. They are usually found some distance from the shore on rough ground. The females reach sexual maturity when 8 cm long. They spawn from April to July and their eggs, 0·7 – 0·95 mm in diameter, are pelagic. The larvae are 2·0 mm long at hatching. The adults never exceed a length of 15 cm.

1 ROCK COOK
3 CUCKOO WRASSE, Male

4 BALLAN WRASSE

5 CORKWING WRASSE

2 RAINBOW WRASSE
3A CUCKOO WRASSE, Female

6 GOLDSINNY

PIPE-FISH AND SEA HORSE

The fish on this page belong to the order Gasterosteiformes, small fish with body encased in bony armour, and the family Syngnathidae whose members have a small mouth at the end of a tubiform snout, jointed bony rings encircling the body, a reduction in the number of fins, and the males carry the young embedded in a fold of skin or a pouch. Generally to be found amongst eel grass, they are well concealed by their vertical posture. The pipe-fish and sea horse live almost entirely on small crustaceans, the prey being sucked into the mouth.

1 **Hippocampus ramulosus** (Sea Horse) appears occasionally in the English Channel and rarely in Irish waters and the North Sea. It lives in shallow water amongst seaweed to which it anchors itself by the tail. It swims in the vertical position; the dorsal and pectoral fins oscillate at rates of up to 70 per second. This enables it to manoeuvre slowly but precisely close to prey. It eats crustacea small enough to be sucked into the tiny mouth, including copepods, isopods and amphipods. Spawning occurs in late spring and summer. A pair observed in an aquarium swam slowly together occasionally snapping their heads while producing quite loud clicking sounds which were continuous during the embrace. The pouch of the male (1A) becomes very large and a placenta-like lining develops in which the eggs are embedded during the 4 – 5 weeks' incubation. As many as 142 young may be carried in the pouch. The young are expelled by the male flexing his body and they look like miniature adults (1B). Shortly afterwards they surface to fill the swim-bladder with air. Adults can reach a length of 15 cm.

2 **Entelurus aequoreus** (Snake or Ocean Pipe-fish) can be found all around the British Isles at depths of 30 – 100 m. In summer it comes to the surface and attaches itself by the tail to seaweed. Spawning takes place in deep water from mid-May till the beginning of October. The eggs are attached to the glandular surface of the male's abdomen and incubate for 4 weeks. The young, about 12 mm long, liberated in a larval condition have pectoral fins and a yolk sac. The larvae and the post-larvae remain in inshore waters at depths of about 14 m. Adults may reach a length of 61 cm.

3 **Syngnathus (Siphostoma) typhle** (Broad-nosed Pipe-fish) is a fish of the eastern Atlantic found occasionally around the British Isles. It remains almost upright among *Zostera*, eel grass, moves only slowly, and is very hard to find.
The spawning period is in May and June when there is a migration into water 4 m or more in depth. The male's brood pouch is formed by two folds of skin which extend along the underside of the tail. The male receives eggs from one or more females and fertilizes them. The folds of the pouch now come together and the lining thickens, processes develop and grow between the eggs. A network of fine blood vessels conveys oxygen to the developing embryos, and nutrient is secreted. After 3 – 4 weeks the young miniature adults, 20 – 25 mm long, are released and the pouch lining is shed. The young soon begin to feed but may remain near the male. For two months they are planktonic often over quite deep water. Pipe-fish manoeuvre themselves close to small prey, which they suck in; crustacea such as mysids and amphipods are eaten as well as young fish. This species can reach a length of 30 cm, and males are mature when 12 cm long.

4 **Syngnathus acus** (Greater Pipe-fish) is found along the coast in deep water where the bottom is of rocks and weeds. It is common all around the British Isles. The spawning period lasts from May to August with a peak period in June and July. The eggs are large, 2·5 mm, and develop in the brood pouch of the male for about 5 weeks. The young at emergence are 22 mm long and remain in the plankton during the summer, sometimes being found far from the shore. Adults may be 46 cm long.

5 **Nerophis ophidion** (Straight-nosed Pipe-fish) is found from Algiers in the Mediterranean to Norway. Spawning takes place early in summer. The young, liberated when about 9 mm long, are (like 2 and 7) provided with pectoral fins and are pelagic. Adult females reach a length of 30 cm, males 20 cm. Food is almost entirely crustacean larvae.

6 **Syngnathus rostellatus** (Nilsson's Pipe-fish or Lesser Pipefish) occurs in the English Channel, the North Sea, around Ireland and along the west coast of England. It lives amongst floating or attached seaweeds where the bottom is of sand, or in brackish water. Spawning takes place in summer and the eggs are protected by the male in his brood pouch. The young, 1·4 cm long when they emerge, are planktonic. Adults reach a length of 17 cm.

7 **Nerophis lumbriciformis** (Worm Pipe-fish) can be found from the English Channel to Norway. It lives beneath stones or amongst seaweed in rock pools between the tide marks, particularly during summer. In winter it may go into deeper water. The spawning season begins late in April and continues until July, as males with eggs attached to the glandular surface of the abdomen have been caught then. The larvae, about 9 mm long when released, are transparent and have pectoral fins. They are pelagic, remaining near the surface for several months during summer and autumn, changing gradually into adults. A length of 15 cm may be reached.

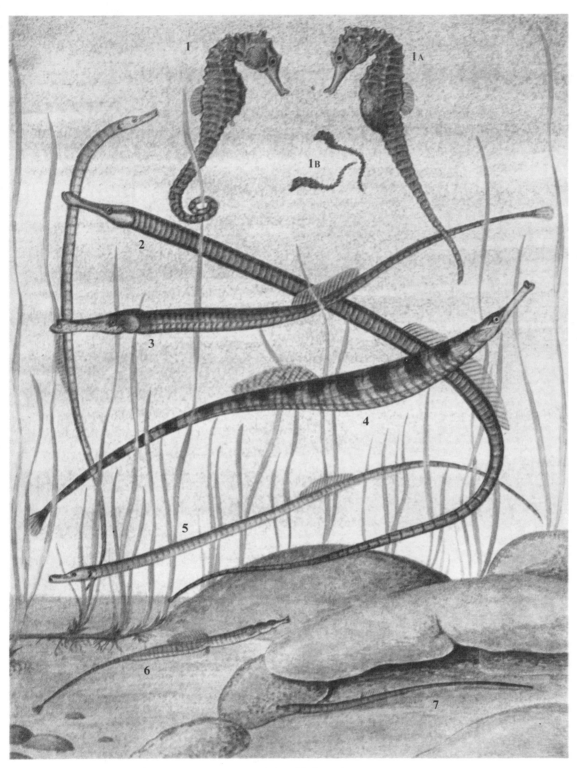

1 SEA HORSE. Female   1A SEA HORSE. Male with brood pouch   1B SEA HORSE, young
2 SNAKE PIPE-FISH   3 BROAD-NOSED PIPE-FISH
4 GREATER PIPE-FISH   5 STRAIGHT-NOSED PIPE-FISH
6 NILSSON'S PIPE-FISH   7 WORM PIPE-FISH

BULLHEADS

The fish on this page belong to the order Scorpaeniformes (p. 40) and are members of the family Cottidae. The bullheads have a broad flattened head with one or more prominent spines on the gill cover. The body is without scales but may have small prickles or embedded bony structures each with a small spine. None has a swim bladder. Some species are capable of producing growls particularly when feeding. All are small fish living in shallow seas or in fresh water. They display sexual dimorphism, the genital papilla being long and pointed in the male of most species, but short and round in the female. There is usually an elaborate spawning behaviour. The eggs are demersal, and generally adhere to each other, being deposited in crevices or beneath stones. The larvae are planktonic.

1 **Myoxocephalus (Cottus) scorpius** (Short-spined Sea Scorpion) is widely distributed along the coasts of the north Atlantic, and is common on all British coasts except the English Channel. It is generally found between tide marks, the exception again being the Channel where it lives in water 4 – 60 m deep.

Sexual maturity is reached in the second year when 98 mm or more in length. Spawning occurs in January off the west coast of Ireland whereas eggs are found in April around the Isle of Man. The eggs, 1·5 – 2 mm in diameter, have several oil globules and are orange in colour. A number are laid in a few seconds in a clump, usually in a hole or cleft in the rocks or attached to a stone. The capsule is thick and bears small processes by means of which it adheres to surrounding eggs. The male probably guards and fans the eggs during their long incubation period of some 5 weeks.

When newly hatched the larvae, 7 – 7·5 mm long, are olive-green in colour with a series of black chromatophores over the head and at the base of the pectoral fin where there are areas of yellow pigmentation. The young fish become pelagic shortly after the yolk has been utilized and remain so until about 20 mm long. They then come to inhabit shallow water and have by this time acquired adult characteristics. The adults feed mainly on crustacea and fish. This, the largest of the sea scorpions in British waters, can reach a length of 30 cm. The record rod-caught specimen weighed 907 g.

2 **Taurulus (Cottus) bubalis** (Long-spined Sea Scorpion or Sea Scorpion) is a fish of the north-east Atlantic, found from the Bay of Biscay to Norway and into the Baltic Sea. It is common around British coasts where it inhabits rock pools, often the same one for quite long periods. Occasionally it will enter estuaries.

They usually attain sexual maturity at the end of the first year. The spawning period varies considerably depending on the area. Off the west coast of Ireland it lasts from January to April, off Plymouth from February to June, in Manx waters from March to April, and further north from March until May. There are two spawnings per year. Each egg is 1·7 – 1·88 mm in diameter and is demersal.

There is an elaborate courtship during which the male, with enhanced colours, displays before the female. With fins erected he makes jerky movements occasionally even engulfing the head of the female in his mouth. The eggs are laid in a clump, attached to the substratum, and are subsequently fertilized by the male who then guards and fans them until they hatch in 42 – 50 days. At hatching the young are about 4·5 mm long and are planktonic, remaining so until 10 mm in length. The food of these small fish, 4·5 – 10 mm long, consists of crustacea, such as *Temora* and the nauplius larva of *Balanus*, and of diatoms and algae. The post larvae are usually found below 20 m. The adults are predators, swallowing their prey whole. They feed mainly upon crustacea, including amphipods, decapods, isopods, and mysids, but also take fish including gobies, sand eels, and their own species. The intensity of feeding usually falls after the breeding season to a minimum in April and May, and then rises to a peak in the warmer months. A second fall from September to November is followed by a rise to another peak before the breeding period. The maximum length reached by this species is 17 cm and they have a lifespan of 3 – 4 years.

3 **Taurulus (Cottus) lilljeborgi** (Norway Bullhead) has been recorded from the Firth of Clyde, the Isle of Man, the Irish Sea, and off Scotland. It has been found at depths of 18 – 87 m over coarse ground.

This species may have a wider distribution than previously realized because of confusion with *T. bubalis*. It can be distinguished from the latter by the presence of two longitudinal rows of ossicles (small bones) dorsal and parallel with the lateral line, the presence of 12 fin rays compared with 13, and a maximum length of 6 cm instead of 17 cm.

1 SHORT-SPINED SEA SCORPION
2 LONG-SPINED SEA SCORPION
3 NORWAY BULLHEAD

GURNARDS

The fish on this page belong to the order Scorpaeniformes (p. 40) and the family Triglidae. All have wedge-shaped heads protected by strong bony plates and spines. The thin tapering body is covered with scales, the head is scaleless. The last 2 – 3 rays of the pectoral fin are separated to form feeler-like processes, used to explore the bottom for prey and to 'walk'. Some species can swim fairly fast, 129 cm per second. A very muscular swim bladder is present from which originate loud grunts and groans, probably for communication purposes.

1 **Trigla lucerna (hirundo)** (Gurnard, Yellow Gurnard, Tubfish) has a distribution from the Mediterranean to Norway. It is common all around Ireland and along the Channel coasts penetrating the southern North Sea, and is occasionally found in Manx waters. It inhabits sandy or mixed bottoms in shallow water or in deep water beyond the continental shelf. The female reaches sexual maturity when 32 cm long. The extent of the spawning season is not known although spent fish have been found in August. The eggs are 0·8 mm in diameter. Young *T. lucerna* are often called Sapphirine gurnards because their large pectoral fins have a bright blue margin and a round dark blue area with a pattern of pale spots at the base. The colours are lost as the fish grows larger. Food consists of crustaceans, molluscs, and slow-moving fish and worms. It is the largest of our common gurnards, reaching almost 60 cm in length. The record rod-caught specimen weighed 5·195 kg.

2 **Eutrigla (Trigla) gurnardus** (Grey Gurnard) has a distribution from the Mediterranean to Iceland, the north of Norway, and into the Baltic Sea. It is fairly common around the British Isles, usually at depths of 2 – 160 m on sandy bottoms. In spring and summer an annual migration inshore occurs to sandy bays where it is often found between the tidemarks. The males reach sexual maturity when at least 18 cm long and 2 – 3 years of age, and the females when at least 23 cm and 3 – 4 years. Spawning occurs from January to August in the western Channel, February to June in Manx waters, and April to August in Irish waters. There is a gradual process of ripening of the eggs. A large number are produced; one female, 34 cm long, may bear 297,000. The adults spawn in inshore waters and afterwards migrate seawards the females outnumbering the males. The egg is pelagic, 1·2 – 1·52 mm in diameter, with several oil globules which later coalesce. They hatch in 8 – 14 days, depending on the temperature. At hatching the larva is 3·5 – 3·8 mm long, open-mouthed and has a large yolk sac. The second day the pectoral fins increase greatly in size. By the time the larva is 4 mm long these develop into a pair of large efficient fanlike paddles projecting at right angles from the body. The larva is now very active. The mouth still gapes widely. Larvae 5·8 mm long are no longer planktonic but live on the bottom in fairly still water about 50 m deep. The pectoral fins reach nearly to the base of the tail when the

little fish are 15 mm long and at a length of 20 mm the last three rays of these fins have become separated. Most other adult features have also developed. After 1 year the average length is 10 cm, after 2 years 16·5 cm, after 3 years 22·8 cm, the maximum 45 cm. The chief food is crustacea, particularly shrimps, but sand eels, gobies, herring fry and small flatfish are also eaten. It is often caught at night near the surface by anglers, although it lives mainly on the sea floor, occasionally swimming in mid-water. The record rod-caught specimen weighed 623 g.

3 **Aspitrigla (Trigla) cuculus** (Red Gurnard) can be found from the Mediterranean to Norway and is common all around the British coasts except the north-east. It inhabits greater depths than our other species, generally 20 – 40 m but down to 330 m, where the bottom is of sand or sandy mud. They spawn from January to June. The eggs, 1·47 – 1·61 mm in diameter, have a copper-coloured oil globule. At hatching the larvae are 3·7 mm long. The larvae and post larvae can be caught at a depth of 40 m. Adults can reach a length of 39·6 cm but are usually 20 – 33 cm. Their food consists of crustacea, and fish such as dragonets and young flatfish. The record rod-caught specimen weighed 1·417 kg.

4 **Trigloporus lastoviza (Trigla lineata)** (Streaked Gurnard) can be found from the Canaries and the Mediterranean to the south-west coasts of the British Isles. It has been taken off the south coast of Ireland, off Plymouth and, rarely, in Manx waters. The adult reaches a length of about 35 cm.

Trigla lyra (Piper) occurs occasionally off the south-west coasts of England and Ireland. It is a stoutly built gurnard, red without markings. It has a deeply forked snout and large spines behind the gill cover. The lateral line is smooth and the sides are not ribbed.

Aspitrigla (Trigla) obscura (Long-finned Gurnard). Found in the Mediterranean, it occasionally reaches the south-west of the British Isles. A slender fish with a small head, it is characterized by the very long second spine of the dorsal fin. Along the lateral line is a series of leaf-like plates. Red above, it has a silver band along each side and is white below. A length of about 30 cm can be attained.

1 YELLOW GURNARD   2 GREY GURNARD
3 RED GURNARD   4 STREAKED GURNARD

DRAGONETS

The fish on this page belong to the order Perciformes (p. 24) and the suborder Callionymoidei. The males, in their brilliant courtship colours, well deserve the generic name *Callionymus* (Greek 'famed for beauty'). In contrast the females are inconspicuous and well camouflaged. All are small fish and live on the bottom. They show some dorso-ventral flattening and their bodies are without scales.

1 **Callionymus maculatus** (Spotted Dragonet) has a distribution from the Mediterranean to the adjacent Atlantic from where it occasionally visits the west coasts of the British Isles. It will inhabit deeper water than *C. lyra* but also lives in shallow areas where the bottom is of sand. It feeds on small bottom-living animals including worms and crustacea. The spawning season lasts from June to August and the eggs, 0·7 mm in diameter, are ridged as in *C. lyra* and like them are pelagic. Post larvae are found from June to September off the south-west coast of England. The largest specimen taken at Plymouth was 12·4 cm long but the males may reach a length of 14 cm and the females 11 cm.

2 **Callionymus reticulatus** (Reticulated Dragonet) is found in the western Mediterranean, the adjacent Atlantic and northwards to Ireland, the Isle of Man, the English Channel and the southern North Sea. This species, first reported off the British coasts in 1951, is the least common of our dragonets and has a localized distribution.
The males can attain a length of 108 mm and the females 65 mm; they reach sexual maturity when 72 mm and 59 mm long respectively. The secondary sex characters of the males develop gradually, giving them a different appearance from the immature males and the females. The second dorsal fin gradually develops three dark spots on the membrane between the consecutive rays, the uppermost spot being very faint. The spawning period lasts from March until June. The eggs are 0·5 mm in diameter.

3 **Callionymus lyra** (Common Dragonet) is the most common of the dragonets found around the British Isles. It lives in shallow water and down to depths of 50 – 100 m and it feeds upon crustacea, molluscs, worms, and other bottom-living animals.
The male is one of our most beautiful fish when it develops sexually and acquires the bright courtship colours in spring and early summer. It usually matures when 16 cm or more in length as also does the female.

The male is shown in display attitude in the illustration opposite. The spawning season begins in January and continues until August. Their courtship behaviour may be seen in aquaria and was first described almost a hundred years ago. It was said then that the very long dorsal fin of the males was used 'for the purpose of fascinating their mates'. Both fish become excited, particularly the male who darts over the bottom while exhibiting his finery. The dorsal fins are erect, the mouth protruded, the gill covers inflated, and the pelvic fins held rigidly forward and outward. He remains on the bottom, the anal fin depressed and out of sight. The colours are much enhanced in the first few days of sexual activity. The male approaches a female and, if she is ripe and accepts him, they swim together slowly along the bottom. When close the male raises the front of his body off the bottom while the female takes up a position between his gill covers, pectoral and pelvic fins. Next, the male takes the female upwards until they are in a vertical position and well clear of the bottom. He then inclines part of his trunk towards the female, with the genital papilla directed into the funnel formed by the two fish. Along this funnel the products of the gonads are expelled. It seems probable that the male spawns only once and subsequently dies as all the spent males disappear suddenly after June or July. During the breeding season the male loses much weight. The eggs, 0·7 – 0·95 mm in diameter, can be distinguished from other pelagic eggs as they have a series of fine ridges, giving a honeycomb-like appearance. There is no oil globule present. Hatching takes 14 days and the larvae are 2 mm long when they emerge. Larvae and post larvae can be caught off Plymouth from February to December, with a peak in May, and from March until September in Irish waters. They are usually caught at a depth of 10 m or more. Young dragonets spend their first year close to the shore and will even ascend 5 – 6 km up river.
The males grow faster than the females, and are larger, reaching a length of 23·9 cm and a weight of 120 g when 5 years of age. The females do not exceed a length of 18·5 cm but may live for some 7 years.

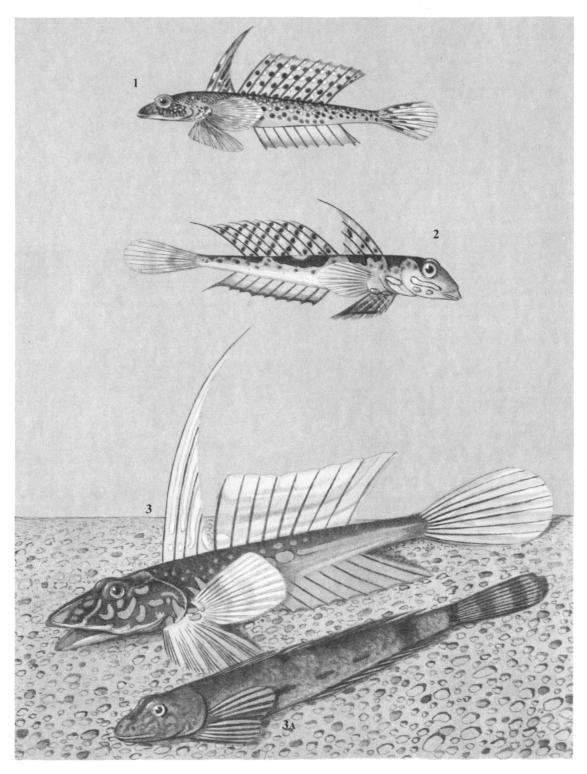

1 SPOTTED DRAGONET. Male   2 RETICULATED DRAGONET. Male
3 COMMON DRAGONET. Male, courtship colours and display 3A COMMON DRAGONET. Female

BLENNIES

The fish on this page all belong to the order Perciformes (p. 24) and the suborder Blenniodei, small fish inhabiting shallow, rocky, inshore waters, that have an elongated body, rather slimy skin which lacks scales, and a mouth generally large and terminal with small, sharp, close-set teeth.

1 **Blennius gattorugine** (Tompot Blenny), common on rocky shores off the south-west coasts, is usually only uncovered in pools at very low spring tides. The spawning period is in the spring off Plymouth, and from May to July off the west coast of Ireland. The eggs, which take 4 – 5 weeks to hatch, are guarded by the male who shows aggressive behaviour to intruders. Newly hatched larvae are 4 mm long and very active. Larvae and post larvae 5 – 8 mm long can be caught from May until August.

2 **Chirolophis ascanii** (Yarrell's Blenny) is present in Irish waters, the English Channel, and Manx waters. It is commonly found on mussel beds where the lamellibranch mollusc *Modiolus* is present in large numbers, in water 40 – 66 m in depth. The spawning period is from October to November and the larvae are pelagic. Post larvae are abundant off Plymouth from January to April with a peak in March.

3 **Coryphoblennius galerita** (**Blennius montagui, B. galerita**) (Montagu's Blenny) is occasionally found off the south-west coasts. It lives in pools and feeds mainly on barnacles. Like other shore fish it is subjected to a tidal cycle twice each day. This is reflected in two peaks of swimming activity that are thought to be related to feeding. In darkness it is less active. It may reach a length of 8 cm.

4 **Blennius pholis** (Shanny), probably the most common of the British shore fish, usually inhabits the lower half of the intertidal zone of rocky shores which offer suitable shelter. It has a distribution from the Mediterranean and the adjacent Atlantic coasts into the southern North Sea.
They mature when nearly 2 years old and spawn when 8 cm long. They breed from April to mid-August, the male remaining ripe longer than the female. Each individual spawns successively — one female may spawn three times in a season. The eggs, 1·42 mm in diameter, are abundant on the shore from mid-April to mid-August, with a peak in June. Spawning behaviour is elaborate. The male hides beneath a stone around which he displays territorial behaviour. In the posture of threat the dorsal, anal and caudal fins are spread, the head and anterior part of the body are raised, the floor of the mouth lowered, and the colour darkens. Another fish may take up a submissive posture lying with its back

to the dominant fish, with fins folded, and pale in colour.
During breeding and until the eggs hatch there is a change in colour from dark brown with black patches to a sooty black which enhances the whiteness of the fold around the jaws. The pair lie side by side in the nest, then the female turns on her side or back so that the ventral surface can be applied to the stone. Her tail quivers, the pectoral fins move slowly and the mouth remains slightly open while each egg is deposited so that the attachment disc faces the stone surface. She moves so as to distribute them evenly over several square cm. The whole procedure lasts 7 – 8 hours. During this time the male periodically becomes excited, and moves over the eggs to release milt to fertilize them. He guards the nest for the incubation period of 43 – 61 days, at 9 – 15°C. He maintains a current of water, by means of tail movements, to aerate the eggs, and brushes them with his body to keep them free from fungal growths. At hatching, the larvae, about 5 mm in length, have already utilized the yolk and have teeth. Young fish 5 – 6·5 mm long can be found from June to September. At one year they are 7 cm, at 2 years 9·5 cm, at 5 years 14 cm long. They may live for 10 years and reach a length of 16 cm. They inhabit rock pools. The home pool is used as a base from which excursions are made at high tide. Their food consists of common shore invertebrates and plants and their intake increases between July and November. Crustacea, polychaetes and molluscs are eaten but their chief food throughout the year is barnacles.

5 **Blennius ocellaris** (Butterfly Blenny) can be found on south-west coasts, usually at depths of 30 – 50 m over coarse gravel. They spawn in spring and summer, the eggs being laid in empty whelk shells, in water 30 – 40 m deep. At hatching the larvae, 4 – 6 mm long, can maintain themselves against small currents and can make small leaps away from danger. This blenny may reach a length of 20 cm.

6 **Lumpenus lumpretaeformis** (Snake Blenny), an inhabitant of the north Atlantic, is found in Irish and Manx waters. They migrate during winter into the North Sea where spawning occurs; larvae have been caught there in March. The largest one caught in Irish waters was 21·2 cm long, while one of 15·9 cm weighed 5 g.

1 TOMPOT BLENNY
3 MONTAGU'S BLENNY
5 BUTTERFLY BLENNY

2 YARRELL'S BLENNY
4 SHANNY
6 SNAKE BLENNY

PERCH-LIKE FISH OF COASTAL WATERS

The fish on this page belong to the order Perciformes (p. 24) except 4. The suborder Mugiloidei is represented by 1, 2 and *Liza ramada*. These fish are distinguished by the two small but widely separated dorsal fins, the first of which is supported by spines. The pelvic fins have a spine and five rays and are some distance behind the pectoral fins. They are found in shoals in coastal areas, usually in bays which have freshwater running into them. The butterfish belongs to the suborder Blennioidei (p. 72), fish with very small scales and a long dorsal fin. The sand smelt is a schooling fish of shallow waters that has a terminal, upturned mouth and belongs to the order Atheriniformes (p. 32).

1 **Crenimugil (Mugil) labrosus** (Thick-lipped Grey Mullet) has a distribution from the Canaries along the eastern coast of the Atlantic northwards to Norway. It is found all around the British coasts, and is the only mullet in Irish waters. Large shoals may appear in spring in sandy bays or estuaries, but disappear in autumn, probably to winter in deeper water, except off the English south coast where they often remain in sheltered harbours, particularly where warm water from industrial plants enters the sea.

The smallest mature male found was 35 cm long and 9 years, the smallest female 38 cm and 11 years. Spawning lasts from January to April and one known area is off the Scilly Isles. The eggs, 1·29 mm in diameter, have a cluster of oil globules. Females 47 and 56 cm long produced 372,000 and 685,000 eggs respectively, shed in several spawning acts. Young fish 1 cm long appear in coastal waters in May. By summer silvery ones of 2·5 cm can be found in shore pools swimming and darting at the surface.

During feeding the fish takes up a vertical position with its mouth on the bottom. Three processes can be seen: sucking, scraping, and straining. The fish sucks food and mud particles into the mouth, expels a strong jet of water from the gills that stirs up mud, then ejects mud from the mouth. It scrapes algae and diatoms from vertical surfaces as it moves up and down them. Finally it goes into a cloud of mud which it has stirred up and filters it through the gill rakers. The food is mainly diatoms and algae, but also crustacea including copepods, ostracods and amphipods, as well as eggs and worms. These fish defend their feeding territory against intruders of the same species. They possess a long gut — 240 cm in a fish of 50 cm and the food, much of it indigestible, takes several hours to pass through. Growth is quite slow; a fish at 2 years is 9 cm in length, at 4 years 24·6 cm, at 8 years 38·6 cm. Fish of up to 60 cm have been caught and one of 58·5 cm weighed 2·338 kg. They live for 17 years or more. The record rod-caught specimen weighed 4·564 kg.

2 **Liża (Mugil) auratus** (Golden Grey Mullet) is common in the Mediterranean and in the eastern Atlantic from the Canaries to southern Norway. A few are found in estuaries off the Devon coast, the Isles of Scilly, and the Channel Isles where the record rod-caught specimen of 779 g was captured. Post-larvae, 6 – 27 mm long, have been caught off the Devon coast. Growth is fairly slow; fish of 2 years are 10·7 cm long, at 4 years 24·2 cm, and 6 years 34·3 cm. A length of 41 cm can be attained.

3 **Pholis gunnellus** (Butterfish, Gunnel) is a north Atlantic fish which reaches as far south as the Channel. It lives between tide marks on rocky shores, or in the region of low water spring tides amongst oarweed, usually under stones. Sexual maturity is reached in 2 years when 9 – 10 cm long. The spawning season is in February in the western Channel, and January to March off north Wales. The eggs, 2 mm in diameter, are deposited in small masses, often in rock crevices, and guarded by both parents initially, but later mainly by the female. After probably 6 weeks they hatch, and the larvae, 5 – 9 mm long, lead a pelagic life for some months drifting with the currents and the tide. At 3 years they are 13 cm long. They live for 5 or more years attaining 16 cm in length. They feed seasonally, eating more during the warmer months, on polychaetes, molluscs, and crustacea including amphipods, isopods, decapods and barnacles.

4 **Atherina presbyter** (Sand Smelt) can be found in the Mediterranean, the adjacent Atlantic, all around Ireland, in the English Channel and the southern North Sea. An inhabitant of *Zostera* (eel grass) beds in bays, harbours or estuaries, it feeds on the larvae of crabs or mysids as well as the fry of mullet and small sand eels. Fish 9·5 cm long are mature sexually. Spawning occurs in shallow water in June and July and the eggs, 1·85 – 1·9 mm in diameter, have filaments which become entangled with the seaweed. The newly hatched larva, 6·8 – 7·5 mm long, is transparent. The early life is spent inshore and by August or September it is 25 – 45 mm long. Adults can attain a length of about 17 cm.

Liza ramada (Mugil capito) (Thin-lipped Grey Mullet) has been captured in estuaries along the south and south-west coasts of England. It sometimes goes up river into freshwater. The spawning habits are unknown but two females with ripening ova were caught in August and September; one female 53 cm long, 1·450 kg in weight, contained 19,736 eggs.

1 THICK-LIPPED GREY MULLET   2 GOLDEN GREY MULLET
3 BUTTERFISH   3A BUTTERFISH.Male, guarding eggs laid in bivalve shell
     4 SAND SMELT

ROCKLINGS

The fish on this page belong to the order Gadiformes and 1–5 to the family Gadidae (p. 36). They have a large mouth with fleshy lips and sharp teeth. There is a single barbel beneath the chin and near each nostril there is one or more pairs of sensory filaments or 'beards'. These fish are largely nocturnal and seek their food by smell and touch. They swim along the bottom with the tip of the chin barbel and the pelvic fin touching the substratum.

1 **Ciliata (Onos) mustela** (Five-bearded Rockling) can be found all around the British coasts. It has a distribution from the Mediterranean and the adjacent Atlantic coasts, northwards to northern Norway and Iceland. It is common in intertidal rock pools near the low water mark where it may be found beneath stones.

The females are mature sexually when 5 cm long and the males at 6 cm. The spawning season is from January to August in Irish waters and to July off Plymouth. The eggs, 0·6 – 0·9 mm in diameter, hatch in 5 – 10 days, depending upon the temperature. The larvae at hatching are 2 – 2·5 mm long. The pelagic fry have greatly enlarged pelvic fins tipped with black, and during this phase the first dorsal fin develops. At the beginning of this stage the fry are transparent but gradually the body acquires a silvery appearance. In the summer the larvae are often present in coastal waters where they spend their first year of life. Demersal post larvae, 3·9 cm or more in length, are found in shallow sandy bays. These early adolescents are often called 'Mackerel midges', a phase when the back is green-blue, the sides and belly silver, and the fins rather long. Metamorphosis to the dull adult colouration takes place when they are 5 cm or more in length. At this size they move into deeper water, 50 – 60 m, often over rough ground where there are mussel (*Modiolus*) beds.

The food of this species consists of crustacea such as amphipods and isopods, and also molluscs and small fish. Adults may reach a length of 21·5 cm. The record rod-caught specimen weighed 262 g.

2 **Rhinonemus (Onos, Enchelyopus) cimbrius** (Four-bearded Rockling) is found all along the Atlantic coasts reaching as far south as Cornwall. It generally lives in off-shore waters, 60 – 100 m deep, over mud or muddy sand.

The spawning season lasts from April to September. The eggs, 0·66 – 0·98 mm in diameter, are pelagic and have a single oil globule. When newly hatched the larva is 2·75 mm long and transparent. Post larvae can often be found on the surface from May to August, 2 – 5 mm long, with elongated, black-tipped pelvic fins. The barbel and sensory filaments gradually develop and the fry becomes silvery. When about 4 cm in length the adult characters begin to appear and they become demersal.

Their food consists mainly of crustacea, particularly *Crangon* and amphipods, but also others such as mysids and *Pandalus*. The adults can reach a length of 25 cm.

3 **Gaidropsarus mediterraneus** (Darker Three-bearded Rockling) is found along the south-west coasts of England and Ireland and in the Irish Sea. It inhabits rocky areas of the littoral and sub-littoral region in water 10 – 20 m deep. It may be found in pools at low water under rocks or stones. The adults can reach a length of 50 cm.

4 **Gaidropsarus vulgaris** (Three-bearded Spotted Rockling) is an off-shore species and not usually found in the littoral zone. The spawning period is from April to June. Adults can reach a length of 55 cm but are sexually mature when about 20 cm. The record rod-caught specimen weighed 1·278 kg.

5 **Ciliata septentrionalis** (Northern Rockling) has only rarely been recorded from the North Sea and off the Atlantic coasts. It has been caught at depths of 30 – 90 m, usually over mussel (*Modiolus*) beds or coarse gravel and sand. The spawning period is in March and April. Its food consists chiefly of decapod crustacea but mysids are also eaten.

Zoarces viviparus (Viviparous Blenny) is a member of the order Gadiformes (p. 36) and the family Zoarcidae, fish with rather elongated bodies and a single dorsal fin which is continuous with the tail and anal fins. The viviparous blenny is found in water 4 – 10 m deep along the Scottish coast, often under stones or almost buried in mud. This is one of only a few teleosts in which fertilization is internal as is the development of the young. At birth the young are expelled looking like miniature adults, 4 – 5 mm long, between December and February. Like the adults they seek shelter amongst stones and weeds.

A length of 32 cm can be reached; the record rod-caught specimen weighed 187 g.

1 FIVE-BEARDED ROCKLING
2 FOUR-BEARDED ROCKLING   3 DARKER THREE-BEARDED ROCKLING
4 THREE-BEARDED SPOTTED ROCKLING   5 NORTHERN ROCKLING

SUCKERS

These small fish of the order Gobiesociformes are highly specialized to adhere to the substratum. They have a soft body without scales or spines, the head is depressed, and the pelvic fins modified to form a sucker. These features and the absence of a swim bladder are adaptations for their mode of life on the bottom in coastal regions.

1 **Lepadogaster lepadogaster (gouanii)** (Cornish Sucker) is common all around Ireland and along the west and south-west coasts of England and Wales. It is found amongst stones in pools near the low water mark.
The spawning period is in spring and summer. The eggs (1A) are yellow and elliptical, 1·37 mm long and 1·08 mm in diameter. They can be found between the tide-marks in a single layer attached to the undersides of stones or in empty bivalve shells. One side of the egg has a flat disc and this is attached to the shell or stone by fine fibrils which are secreted, and which radiate outwards towards the marginal fringe. Both parents guard the eggs during the 4-week incubation period. A newly hatched larva is 4 mm in length. By August little fish of 6 mm are found which have already undergone metamorphosis, have adult form, and can fix themselves to stones by means of their sucker. Adults can reach a length of 8 cm.

2 **Lepadogaster candollei** (Connemara Sucker) occurs locally on the west coast of Ireland, occasionally on the coast of the Isle of Man, and only rarely at Plymouth. It is generally found in rock pools or beneath stones in shallow water near the low water mark.
The spawning period is May to early July on the Irish coast. The eggs are deposited one at a time and incubate for 4 weeks. Larvae 4 – 5·25 mm in length have been caught between June and August, the small ones presumably being quite young. When 6 mm long it is bright red and yellow, with red predominating, and is one of the most brilliantly coloured of the post larval fish. Metamorphosis occurs at this size and takes about 4 days; the adult fins with fin rays form, the ventral sucker develops to a rudimentary state, and the colour changes. The pigment is arranged in regular dorso-ventral stripes of dull purple, with the characteristic star-like black chromatophores still persisting on top of the head. At this size they feed chiefly on copepods and *Balanus* nauplii, which they pursue along the bottom. A length of up to 7·5 cm can be reached.

3 **Apletodon (Lepadogaster) microcephalus** (L. couchi) (Small-headed Sucker). Common around Ireland, the Isle of Man, the Firth of Clyde and Plymouth, it inhabits the lower littoral zone and can be found under stones or amongst seaweed in rock pools. Spawning occurs in June and July, the eggs often being laid in the holdfasts of the seaweed *Saccorhiza*. Late post larvae and juveniles are common at low tide amongst bushy seaweeds, especially *Cystoseira*, during summer and autumn.
The sucker of this species is a double one, each part of which exerts a different suction pressure and provides the fish with an effective means of attachment against strong tidal currents. The flattened head and stream-lined body also help in resisting displacement. A small fish, it can reach a length of 5 cm.

4 **Diplecogaster (Lepadogaster) bimaculata** (Two-spotted Sucker) has a distribution from the Mediterranean northwards to the south-west coast of Norway. It is an inhabitant of coarse grounds outside bays, where it can be found under stones. They are probably sexually mature in their third year when 5 – 5·5 cm in length. The spawning season lasts from May to July or August. Elongated, demersal eggs are laid; these have one of the long sides flattened for attachment to empty shells, often of the scallop *Pecten*. The incubation period lasts for 4 weeks and the larvae, at hatching, are 3 – 4 mm long and pelagic. The yolk is utilized when a length of 5 mm is reached. The sucker becomes visible as a thickening at this size and star-shaped chromatophores are scattered over the head and body. When 7 mm long the dorsal and anal fins are apparent and the head has become flattened dorso-ventrally. Fry of about 6 mm in length are found in quite large numbers during June and July and others in the early stages of metamorphosis. It is able to attach itself to a surface by means of the sucker and changes colour from pink to purple with conspicuous blue spots. The sucker is fully developed when a length of 10 mm is reached and by 17 mm it resembles the adult. It is able to change colour and one fish of 17 mm was bright red when caught but became a dull purple shortly afterwards. In an aquarium these little fish spend much time clinging to the surface of the glass but will detach themselves to dart after food which consists chiefly of small copepods. Adults can reach a length of 5 cm.

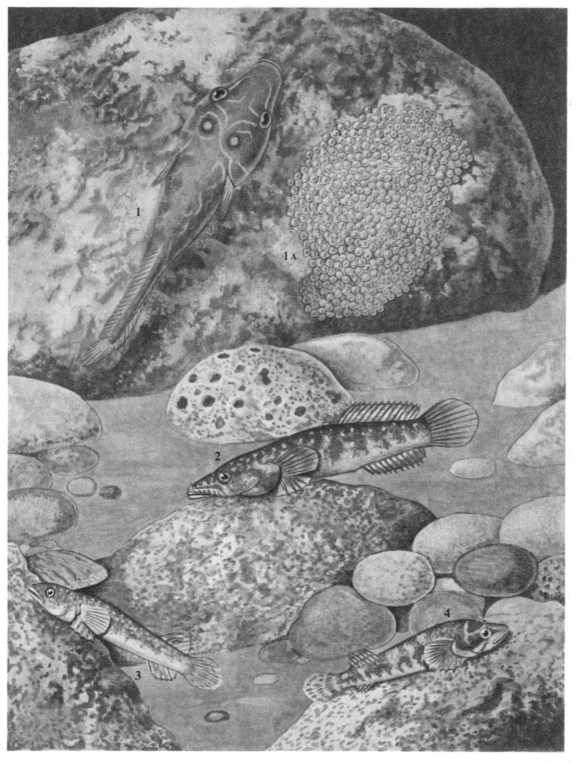

1 CORNISH SUCKER 1A CORNISH SUCKER.eggs
2 CONNEMARA SUCKER
3 SMALL-HEADED SUCKER 4 TWO-SPOTTED SUCKER

MAIL-CHEEKED FISH

These fish belong to the order Scorpaeniformes (p. 40) and all come from northern waters. The sea snails (3, 4) of the family Liparidae, have smooth, naked skins, an almost terminal mouth, small eyes, and a ventral sucker formed by fused pelvic fins. In the lumpsucker (2) too, of the family Cyclopteridae, the pelvic fins have become modified to form a sucker. The body is rounded, the head short and thick, the skin naked but covered with mucus, and the first dorsal fin is replaced by tubercles. The pogge (1), a small fish of the family Agonidae, is entirely covered with bony scales while the head is armed with spiny plates.

1 **Agonus cataphractus** (Pogge) is a fish of the north Atlantic which extends southwards to the English Channel. It is found on gravel at depths of 20 - 40 m and also in estuaries. The prey consists of small crustacea, echinoderms, molluscs and worms. They reach sexual maturity when 8 cm long, and spawning occurs in the winter from late December until mid-February. The demersal egg masses are generally deposited in *Laminaria* holdfasts, in cavities or in rock crevices at the low water mark, but have also been recorded from the middle of the North Sea. The straw-coloured eggs, 1·75 – 1·9 mm in diameter, have a thick capsule; the incubation period probably lasts for 3 months. The larvae are large when they hatch, 6 – 7·5 mm long, and may be caught from February until April, as they are pelagic for the first three or four months. When 19 mm long the adult characters have already developed. Adults can reach a length of 15 cm.

2 **Cyclopterus lumpus** (Lumpsucker, Sea Hen) is found all around the British Isles although less commonly along the Channel coast. During the breeding season, it migrates inshore and inhabits the littoral zone where the young remain for several months before moving off shore. During the winter the adults enter deeper water where there is a vertical distribution according to size, the larger fish going deeper than the smaller ones.
They spawn from January to May and at this time the abdomen of the male is orange-red, that of the female bluish-black. The pink eggs, 2·2 – 2·6 mm in diameter, have a single oil globule. They are laid on the surface of stones or weeds between tide marks. The male guards them for 43 – 60 days, removes foreign particles from them, and aerates them with currents produced by the actions of the pectoral fins and the mouth, and by a reversed action of the gills which becomes very vigorous when the eggs are close to hatching. The male does not feed during the period of incubation; the females take no interest in the eggs after laying them.
The larva at hatching already has a functional sucker by which it may become attached to plants or stones. However, larvae are often carried out to sea and remain pelagic for their first year.

Though adults have been encountered swimming at the surface, they spend much time on the bottom and can descend to considerable depths, having been taken at 300 m. The sucker is an efficient organ; a force of 13·3 kg was needed to overcome the suction pressure of a fish 39 cm long. Even so, it cannot withstand wave action during very heavy seas as is evidenced by the large numbers of lumpsuckers sometimes found on a beach after a storm. Its main function is probably to maintain the adults in the littoral zone during spawning, and especially the male while guarding the eggs.
The lumpsucker's food includes small crustaceans, worms, molluscs, and fish. A male may reach a length of 51 cm and a female 61 cm. The record weight for a rod-caught specimen is 6·435 kg.

3 **Liparis montagui** (Montagu's Sea Snail) is present all around the British Isles. It can be found in estuaries, near the low water mark, the lower littoral zone, and occasionally near beds of the mussel *Modiolus*.
They spawn from January to June, and egg masses are attached to seaweed. Larvae 3 – 6 mm long have been caught between February and September. The smallest larva recorded was 2·7 mm and in specimens of 4 mm the yolk had been absorbed. The larval to post-larval stage lasts about 14 days. Adults may be 8 cm long.
The sucker of this species is efficient and a force of 200 g is needed to break the suction of a fish 8 cm long. A particularly interesting feature is the ability of this fish to increase the suction pressure very rapidly in response to a sudden increase in water speed.

4 **Liparis liparis** (Sea Snail) is a northern fish and consequently is less frequent towards the south but it does extend to the English Channel. It is generally found near the low water mark, under stones or amongst algae, and often in estuaries where the ground is soft. Its food consists mainly of crustacea. They breed from January to March and egg masses are attached to weeds, hydroids, or stones usually below the low water mark and down to 60 m. The adults can reach a length of 15 cm.

1 POGGE

2 LUMPSUCKER. Female   2A LUMPSUCKER. Male   2B LUMPSUCKER. eggs

3 MONTAGU'S SEA SNAIL   4 SEA SNAIL

GOBIES OF COASTAL WATERS

The fish on this page are members of the order Perciformes (p. 24) and the suborder Gobioidei (p. 56).

1 **Gobius niger** (Black Goby) is found in inshore waters around the British Isles, or in estuaries where the bottom is of mud and stones. It feeds mainly on crustacea, but also on molluscs, worms, and small fish. The long spawning period lasts from April throughout the summer. The eggs, deposited close to the shore, are elongated, 1·5 mm in length, and have threads of attachment. The yolk is opaque and lemon-coloured. At hatching the larva is 2·5 – 3 mm long. Larvae and post larvae are found in pools from April to September, with a peak in June and July. When 12 mm long they go to the bottom to live like adults. They probably take 2 years to reach full size, usually about 17 cm.

2 **Chaparrudo (Gobius) flavescens (G. ruthensparri)** (Spotted Goby) is common on the south and west coasts of England and all around Ireland. It lives in weeds in rock pools just below the tidemark and in harbours. Unlike most gobies, it swims in shoals in mid-water. They eat chiefly copepods, but also amphipods and barnacle larvae. They are themselves eaten by coalfish and pollack. Spawning takes place February to August and the ovoid eggs, about 0·8 mm long and 0·6 mm in diameter, are laid on empty mollusc shells near the tidemark. The male guards the eggs for 10 – 16 days, and at hatching the larvae are about 2·6 mm long. Larvae and post larvae, 3·5 – 26 mm, are caught from February till September with a peak in June. Adults can attain a length of 6·4 cm.

3 **Pomatoschistus (Gobius) microps** (Common Goby) occurs all around the British Isles in brackish pools high on the shore, intertidal pools, and estuaries. Crustaceans are its main food. These fish spawn during summer and the eggs, 0·85 – 1 mm long and 0·65 – 0·7 mm in diameter, are attached to empty mollusc shells. After 14 days the larvae, 3 mm or less, hatch. Adults can reach a length of 6 – 7 cm.

4 **Gobius paganellus** (Rock Goby) is found along the west coasts of the British Isles. They are common between the tidemarks especially in rocky areas. Spawning occurs from February throughout spring and summer. The eggs are laid on stones between the tidemarks and larvae hatch after 19 days. The yolk is utilized in 30 days. Larvae and post larvae of 3·5 – 11 mm can be caught from February to July. The adults can reach a length of about 13 cm.

5 **Pomatoschistus (Gobius, Iljinia) pictus** (Painted Goby) is present along the south and west coasts of the British Isles, in inshore waters where the bottom is of coarse sand or shell debris. They spawn from late February until September. The pear-shaped eggs, 0·65 mm in diameter and 0·8 mm long, are attached at the base by threads to empty shells of *Cardium* or *Patella*. After 14 days the larvae, about 3 mm long, hatch. The post larvae may go into water 100 m deep. Adults can reach 5·6 cm length. Although much time is spent on the bottom they will frequently glide up and over the stones. They feed upon amphipods and probably other small crustacea.

6 **Gobius forsteri (Thorogobius ephippiatus)** (Leopard Spotted Goby). This inhabitant of the Mediterranean and adjacent Atlantic, northwards to Iceland, has been found in the western English Channel and on the west coast of Scotland, and was first recorded here in 1956. They are sublittoral and live below the *Laminaria* zone, in crevices near vertical rock faces of the inshore shelf between the low water spring tidemark and depths of 30 – 40 m. They reach sexual maturity in their fifth year at about 8 cm. In an aquarium eggs were laid between May and July. During courtship and incubation of the eggs the male's colour intensifies, the ground colour becoming black-purple. Adults can reach a length of 12·9 cm at 9 years.

7 **Pomatoschistus (Gobius) minutus** (Sand or Common Goby) is present on all coasts of the British Isles. It is common in shallow inshore or brackish waters and in estuaries where the bottom is of mud or sand. They reach maturity about a year after hatching. The spawning period is long; larvae have been found in all months except August and November. The eggs are elongated, 1·1 – 1·4 mm long and 0·7 – 0·8 mm in diameter, with numerous oil globules and a long thread for attachment to empty bivalve shells. Hatching occurs in 14 days and the larva is 3 mm long. When 3·5 mm, after utilizing the yolk, it feeds upon small organisms. Young of up to 6 mm swim at the surface of tide pools with shoals of mysids. Some two weeks later they descend to the bottom. Possibly the adults, 5 – 7·5 cm, do not survive the second winter after spawning. They eat small crustacea, mainly copepods. In rock pools they dart about and then disappear as they come to rest — an illusion due to excellent camouflage.

Gobius cruenatus was recorded in 1970 from the south-west coast of Ireland where it is rare. The adult, up to 16·6 cm long, has a stout black body, It differs from the black goby in having much red in the head region, chiefly the lips.

1 BLACK GOBY

2 SPOTTED GOBY   3 COMMON GOBY
4 ROCK GOBY   5 PAINTED GOBY
6 LEOPARD SPOTTED GOBY   7 SAND GOBY

EELS

The conger and the moray eel as well as the eel on p. 86, belong to the order Anguilliformes, distinguished by their serpent-like form, confluent dorsal and anal fins, and the absence of pelvic fins. The gill openings are small, the skin slimy and either naked or with minute scales embedded in it. They have a large number of vertebrae. The majority of fish belonging to this order are marine and all, without exception, spawn in the sea. Conger eels (1), family Congridae, are large, thick fish with a depressed head and without scales. Moray eels (2), family Muraenidae, are characterized by the absence of both the pectoral and pelvic fins.

1 **Conger conger** (Conger Eel) occurs all around the British Isles. It is abundant off the Irish and Scottish coasts, in the Irish Sea and the English Channel, but is less common in the central southern North Sea. Conger eels were fished by Neolithic man, for their remains have been found during excavations of Shell Mound on the island of Oronsay, off the west coast of Scotland.

The conger inhabits rocky areas, usually from the low water mark down to about 140 m, and has also been caught on the Atlantic slope between 190 and 680 m. It inserts its body amongst the rocks and remains concealed during the day. Divers often see very large congers living in shipwrecks, and quite large ones have been caught in harbours. At dusk it may sometimes be seen swimming at the surface amongst weeds. At night it emerges to hunt for food, which is mainly fish including pollack, rocklings, wrasse, pipefish, hake, soles, plaice and other congers. Cephalopods such as *Octopus*, *Eledone*, *Sepia* and *Loligo* are eaten and also some crustacea.

Females probably begin to mature sexually when about 100 cm in length and reach maturity when 130 cm or more. Spawning takes place in water 3000 m or more in depth, along the whole extent of an area stretching from the Sargasso Sea in the western Atlantic; they also spawn in the Mediterranean. Within this area young larvae, 10 mm long, have been found over great depths. The larvae develop into the leaf-like leptocephalus stage (see p. 87); on metamorphosis there is a great reduction in size, from about 12 cm to 7·5 cm. They now migrate to coastal regions. Young unpigmented specimens are sometimes found in rock pools. The conger does not become so pigmented as the eel at this stage,

and, until about 37 mm in length, it remains pale and pinkish in colour. Only subsequently does it take on the adult colouration. After metamorphosis it feeds and develops to maturity. The gonads of both sexes are large; the female's ovaries are enormous because the eggs all apparently develop at the same time, forming as much as half the total weight of the fish. One fish may have 3 – 8 million eggs. When the gonads begin to ripen, the fish cease to feed. Captive congers were found to undergo considerable degeneration of the tissues at this time, and from this evidence it seems likely that they die after spawning.

The largest conger on record, photographed in 1904, was 280 cm long and weighed 72 kg. The record rod-caught specimen, caught off the Devon coast, weighed 42·097 kg. They are caught commercially, mainly on long lines, off the south and west coasts. They are large strong fish, difficult to kill, and can survive for quite long periods out of water. During very cold weather they may enter deeper water for, although they can tolerate wide temperature changes, a sharp fall will sometimes kill them, and then numbers are washed up dead on the shore.

2 **Muraena helena** (Moray Eel) is a rare visitor from warmer seas. All the records of moray eels are from the last century, and they were caught in the extreme southwest of the English Channel. The last one captured was trawled off Eddystone; it was 113·4 cm long and was a ripe male. Adults can, however, reach a length of 150 cm.

1 CONGER EEL
2 MORAY EEL

THE EEL

This fish belongs to the order Anguilliformes (p. 84) and the family Anguillidae which includes deep sea species. The gill openings are small and narrow and just anterior to the pectoral fins. The skin is exceedingly slimy and has small oblong scales embedded in it. Unlike most of the eels *Anguilla anguilla* spends most of its life in freshwater.

Anguilla anguilla (Eel) has a distribution from the Baltic and Iceland to the Mediterranean. It occurs on all the coasts and enters many rivers and lakes of the British Isles.

The early life of the eel is spent in the sea, and when first described, it was thought to be a distinct species and named *Leptocephalus brevirostris*. In 1896 some of these larvae were caught near Messina, in Sicily, and kept in an aquarium by two Italian naturalists who were fortunate enough to observe the transformation into elvers, animals familiar to them in the rivers but whose early life had previously been unknown. Johannes Schmidt, in 1904, found a leptocephalus larva in the contents of a trawl made in surface waters off the Faroes, and another was caught off the west coast of Ireland. There now followed many years of investigations by Schmidt, that finally revealed the source of the larvae as the Sargasso Sea in the western Atlantic.

The eggs are laid in water 400 – 700 m deep in the Sargasso Sea and from here they gradually rise. At hatching the larva, 5 – 6 mm long, continues to move towards the surface waters drifting eastwards all the time. It feeds upon organisms in the microplankton, that are trapped by its long fragile teeth. As it grows and comes to swim near the surface the larva becomes very thin, transparent, and leaf-like in appearance while the head remains quite small. It is this stage (1A) that retains the name 'leptocephalus' (narrow-headed). It takes about 2·5 years to reach the European coasts. The leptocephalus larva is now about 7·5 cm long and can be caught off the extreme south-west coast of Ireland as early as November, while in April and May it enters rivers along the west coasts of England and Wales. The greater the distances travelled by the larvae, the later is their arrival, and some are 3 years old or more before they enter a river.

As the larva reaches the continental shelf it begins to undergo metamorphosis in which there is a gradual change from the leaf-like form to one more closely resembling an eel. It does not feed at this time and loses weight from 1·5 g to 0·15 g. The body become shorter and rounder, the average length diminishes from 7·5 cm to 6 cm, and the elver (1B), as it is now called, begins to feed again, and increases rapidly in size. Its size continues to increase, but more slowly, throughout its life as a yellow eel. The yellow eel (1C) lives either in a lake or in a river/where it remains for several years. In the day it lurks amongst the stones or remains buried in the mud but it becomes very active at dusk. It hunts mainly by scent and feeds upon crustaceans, molluscs, insect larvae, small fish, and fish eggs. The fish taken include sticklebacks, lampreys, salmonids, flatfish, and most frequently of all, other eels and elvers. The food intake varies, being at its maximum from early spring until summer, and thereafter declining to a low level in the winter when the eel becomes very inactive. The rate of growth varies considerably, not only between different localities, but also between individuals from the same locality. The length of time spent in freshwater probably depends upon many factors, but food availability, temperature, and population size must play some part. The males may remain for 8 – 15 years but the females stay for 10 – 18 or more years. They undergo a final metamorphosis during which the onset of sexual maturation occurs. At this time the digestive organs begin to show degeneration and the eel stops feeding. Adaptation to the changes in salinity, encountered on the downstream, or catadromous, migration, takes place. The pigmentation of the body surface becomes dark on the back and silvery on the sides — hence the name silver eel (1D). There is also development of a retinal pigment characteristic of deep-sea fish. These silver eels migrate downstream and again enter the sea where their fate is almost entirely unknown, although it seems likely that after spawning they die. So far only one adult has been recovered from the open sea, and that was from the stomach of a sperm whale off the Azores. The distance to the breeding grounds is some 5600 km from the coast of Britain, but from the Azores, where *A. anguilla* is common, it is 3200 km. Eels when swimming are capable of reasonable speed, and one 60 cm in length had a speed of 114 cm per second when tested over a short distance. Although the distance to the Sargasso Sea is long, the preparation for this migration has been considerable; the eels have spent many years in freshwater feeding and growing, and they have undergone a second metamorphosis.

There are many problems still to solve about the return journey of the eel to the spawning area. What is the reason for the development of a retinal pigment similar to that of deep sea fish? It suggests that perhaps they swim in deep water. How do they navigate to a particular region in the Atlantic Ocean before spawning from quite different areas in Europe? Why do the males remain in fresh water for a shorter period of time than the females, and become sexually mature at a smaller size? It has been found that the majority of large eels captured are females, and some used to restock a lake without outlets were found to be 140 cm after 50 years. Eels when caught are tenacious of life and indeed can live for quite long periods out of water. They will travel overland, moving with snake-like motion. They are often transported alive to fish shops where they are sold. The record rod-caught eel weighed 3·912 kg.

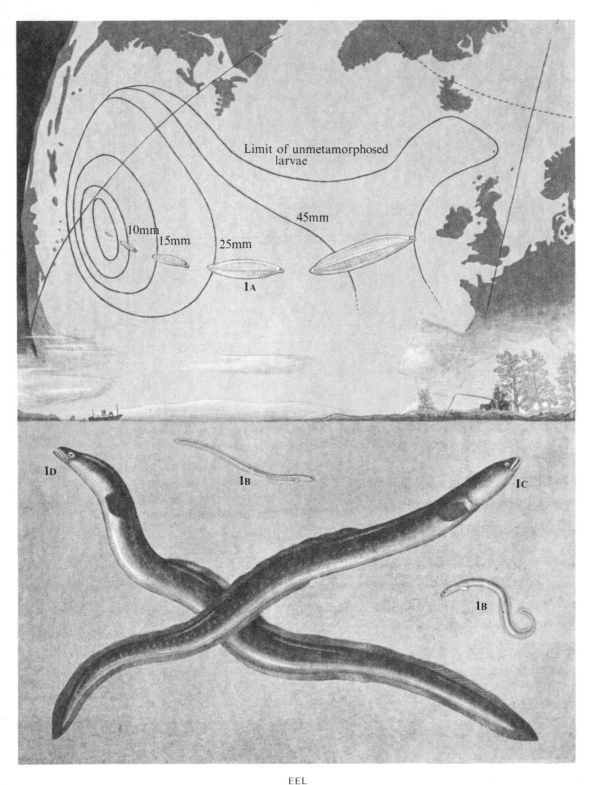

EEL

1A Leptocephalus larvae from eggs laid in Sargasso sea (red lines show size limits)
1B Elvers   1C Yellow eel which swims upstream
1D Silver eel, breeding form which swims downstream to the sea

STURGEON

Two of the fish on this page (1, 2) belong to the infraclass Chondrostei, a rather isolated line descended from a very ancient group of fishes. In the Chondrostei the skull and skeleton are almost entirely cartilaginous, and the body is naked or has vestigial scales. Sturgeons (1, 2) are members of the order Acipenseriformes, primitive fish with an elongated snout and usually 5 rows of conical, bony scutes along the body. The mouth, which is on the underside of the head, resembles a protrusible funnel, and in front of it there are small barbels. The gill chamber is covered by a single gill-cover. A swim bladder is also present. All these fish are anadromous, that is, they ascend rivers to spawn, either from the sea into rivers, or from lower to upper parts of rivers.

Osmerus eperlanus (3), a small inshore fish, belongs to the order Salmoniformes, slender predatory fish of salt and fresh water, and to the suborder Salmonoidei, all the members of which have two dorsal fins the posterior one being adipose.

1 **Acipenser ruthenus** (Sterlet) is a small species usually not more than 52·5 cm long. It lives in freshwater and has been introduced into different countries.

It will only spawn when the water temperature is 30°C. At hatching the young are 7 mm long. When they are 10 weeks they are 50 mm in length and feed mainly on insect larvae.

There is some suggestion of its presence in the Tyne-Tees area.

2 **Acipenser sturio** (Sturgeon) is a fish of the north Atlantic, and during the breeding season it ascends the rivers of North America and Europe. In Europe it can be found from Scandinavia to the Black Sea. When it occurs in the British Isles, it is generally as a solitary straggler, sometimes being caught in coastal waters and occasionally in rivers. In 1954 and 1955 a number were caught in Scottish waters, some were taken off the west coast and others in the North Sea; and some in the English Channel. Since that time, fewer have been captured. Of those recorded in detail the largest was 337 cm and weighed 317·5 kg, while one of the smaller ones was 94 cm and 11·1 kg. Occasional specimens have been recorded from Ireland and up to the beginning of the last century it entered the River Thames. In the reign of King Edward II, about 1324, it was decreed a Royal fish and any sturgeon caught belongs in theory to the Sovereign.

It lives on sandy or muddy bottoms along the continental shelf and feeds on molluscs, worms and other small invertebrates. It stirs up the soft sand or mud with its long snout and the four barbels beneath are used in detecting the prey. The mouth is small and the jaws are weak and without teeth. It is a sluggish fish and moves only slowly over the substratum.

During spawning migrations, it enters rivers of continental Europe in shoals. It usually runs up river at night in early spring. Spawning takes place from March to July and the eggs, 2 mm in diameter and adherent, are laid in flat masses on the bed of the river. The larva, about 9·6 mm long at 3 days is 45 mm long at 28 days. The young apparently leave the rivers at the end of the summer and little is known of their life in the sea. It is considered that they probably mature slowly and some species do not produce eggs for 20 years. In this period of growth and development in the sea it can reach a considerable size and may attain a weight of some 50 kg. The largest sturgeon on record, although not of the same species, weighed 1360 kg.

It is during the spawning migrations into rivers that they are caught. In the Caspian and Black Sea areas the sturgeon forms the basis of two important industries: caviar, a gourmet's food, is produced from the roes of this fish and isinglass, used in making glues and jellies, is obtained from the swim bladder.

3 **Osmerus eperlanus** (Smelt, Sparling) is found along the east coast of England and Scotland, and in Ireland where it is abundant in the estuaries of the River Shannon and its tributary, the River Fergus. It lives in inshore waters along the coast but congregates in estuaries in winter and enters rivers in early spring.

The smallest fish found to be sexually mature was 10 cm long but they are usually larger. They generally spawn at the end of their second year and shoals of these young fish appear first. The mature fish move up estuaries into freshwater to spawn from the end of February to early April in the Wash. This migration takes place at night at the surface, from which they retire during the day. The light yellow eggs, about 1 mm in diameter, adhere to stones and hatch in 8 – 27 days. The newly hatched larvae are 5 – 6 mm long and transparent. By August they are 4 – 5 cm long and in October 6 cm.

They feed upon a wide range of fish including herring, sprats, whiting and gobies. Crustacea are also eaten.

It has been known to reach a length of 32 cm, and a weight of 225 g, but it is usually about 20 cm long.

1 STERLET
3 SMELT
2 STURGEON

STICKLEBACKS

The fish on this page belong to the order Gasterosteiformes (p. 64) and the family Gasterosteidae. Fish of this family have isolated dorsal spines, which can be erected and which have a locking mechanism, providing considerable protection against predators. Of these sticklebacks, 3 is entirely marine, 2 can live in salt or fresh water, and 1 lives in fresh water, being the smallest of our freshwater fish. All exhibit pre-spawning migratory habits and display complex nest-building and breeding behaviour. They are predators and swallow small prey whole.

1 **Pungitius (Pygosteus) pungitius** (Ten-spined Stickleback) is a freshwater fish that shows migratory movements preparatory to spawning. This timid fish constructs its nest amongst dense vegetation — in marked contrast to *Gasterosteus aculeatus* (2) which builds its nest in open areas and is very aggressive. The male after guarding the yellow eggs, 1 mm in diameter, for 14 days until they hatch, constructs a nursery area of loose plant fragments for the young, 6 mm long, to hide in. He cares for them for a few days, after which he may begin another reproductive cycle. Food is mainly small crustaceans, aquatic insects and larvae, and higher crustacea with only a small proportion of other organisms. They are sexually mature when about 34 mm long, and can attain about 55 mm. The ten spines, smaller and less effective than those of the three-spined stickleback (2), still provide some protection against even such large fish as the perch and the pike.

2 **Gasterosteus aculeatus** (Three-spined Stickleback) is common and found in coastal regions, brackish waters, freshwater streams, ponds, and lakes throughout the British Isles.
Sexual maturity can be reached in the first year when they are at least 28 mm long. In spring they migrate upstream to shallow water in shoals, in preparation for breeding in April and May. Each male leaves the shoal and stakes out a territory in open areas. This he will defend against all intruders. He constructs a nest by removing sand, mouthful by mouthful till he has made a depression, into which he piles weeds, preferably algae; after coating them with a sticky substance he shapes them into a mound with his snout. The fish then bores a tunnel, slightly shorter than himself by wriggling through the mound. The sticky material used in making the nest is produced by the kidneys whose function is modified during the breeding season.
The nest completed, the male becomes bright red beneath the chin while the back pales to a bluish white and the eyes are a brilliant blue-green. Now he courts the females, who are swollen with ova and more silvery in colour. He swims in a series of zig-zags whenever a female enters his territory, repeating these movements until a female shows interest by swimming towards him with her head up. Now he swims to the nest and she follows him. He prods the base of her tail to induce her to lay the eggs. Once they are deposited, she leaves the nest and immediately he takes her place and

fertilizes the eggs. Then he chases her from the territory and attracts other ripe females. As many as five females may deposit eggs in the nest, each batch of 50 – 100 eggs being fertilized as they are laid. The male now guards the nest and aerates the eggs; as they develop so he spends more time in ventilating them. They take from 10 – 14 days to incubate depending upon the temperature. The larvae, about 4·5 mm long and transparent at hatching, have a yolk sac, which is utilized in about 8 days. The male cares for the young until they are able to feed themselves; he will pursue stragglers and return them to the nest in his mouth.
Sticklebacks are predators. They stalk their prey mainly by sight, and they swallow it whole. Their food consists principally of small crustacea, aquatic insects and larvae, and oligochaetes. At 1 year sticklebacks can reach lengths of 18 to 48 mm. Subsequent growth is slower, reaching the maximum, 73 mm, at 3 years. They usually live about 3·5 years, maximum 4 years, and in the last few months of life there is a decrease in length.
These fish are caught in large numbers by boys and girls and that they still abound, particularly in ponds, is an indication of their success. Although a relatively small number of eggs is produced, the male's care ensures a good survival rate. Moreover their spines also provide excellent protection against even such large predators as pike.
It is easy to keep them in an aquarium if they are provided with pond weeds and food, and interesting to watch their breeding habits in the spring.

3 **Spinachia spinachia** (Fifteen-spined Stickleback) is confined to salt water and can be found in rock pools, at the margins of rocks, and in harbours during the summer. It is solitary in habit and lives amongst *Zostera*, the eel-grass, so that it is difficult to find. The breeding season lasts from April till June, the male at this time taking on a bluish tinge. After making a nest amongst the seaweed, the male guards the amber coloured eggs, 2 mm in diameter, until they hatch. This takes between 3 and 4 weeks and at hatching the larvae are about 6 mm long. It has been suggested that after breeding the adults die. The young reach a length of 10 – 11 cm in 6 months and at this size can swim with a speed of 72 cm per second. The maximum length attained by this species is 20 cm.

Sea water   Fresh water

1 TEN-SPINED STICKLEBACK

2 THREE-SPINED STICKLEBACK, Female   2A THREE-SPINED STICKLEBACK, Male

2B THREE-SPINED STICKLEBACK, marine form   2C THREE-SPINED STICKLEBACK, nest

3 FIFTEEN-SPINED STICKLEBACK   3A FIFTEEN-SPINED STICKLEBACK, nest and eggs

BURBOT, LOACH AND CATFISH

The fish on this page live in slow-moving water. The super-order Ostariophysi, to which 2 and 3 belong, is the second largest group of primitive bony fish and they live, with few exceptions, in freshwater. Their fins consist mainly of branched and jointed fin rays. They all have cycloid scales except members of the order Siluriformes of which 2 is the only European representative. The skin of this family is slimy and entirely naked, and the mouth broad with 4 – 5 rows of small teeth. A characteristic feature is the barbels. The long mobile pair on the upper jaw reach to the base of the pectoral fins; the other two pairs on the lower jaw are much shorter and have a cartilaginous base. The less specialized members of the super-order Ostariophysi all belong to the order Cypriniformes, mainly small freshwater fish of northern temperate regions. This order includes the family Cobitidae, to which 3 belongs, characterized by a worm-like shape. The small ventral mouth is surrounded by barbels, and the jaws are without teeth.

The burbot (1) belongs to the order Gadiformes and to the family Gadidae (p. 36) of which it is the only freshwater representative. The scales are very small and the thick skin is covered by a slimy secretion.

1 **Lota lota** (Burbot) seems only ever to have been locally abundant and is now found only in eastward flowing rivers from Durham to the river Great Ouse. The earliest English reference to it is a 10th-century one. The largest recorded here were among specimens caught in the 19th century — one weighing 2·5 kg was 69 cm, another of 2·85 kg was not measured. After this time there was a decline until more recent years when a number have been caught and reported. It inhabits deep water in clear brooks and rivers and is not found in enclosed still water in this country although elsewhere it flourishes in lakes. Its chief period of feeding is winter when it becomes very active, especially if conditions are icy. In warm weather it appears to aestivate. Although night is the time when it goes in search of prey, it will also pounce on small fish who venture too close during the day. The eyes are rather small and, like those of other vertebrates who are active at night, have many more rods than cones in the retina. The well-developed olfactory apparatus receives information from the large nasal organs. Spawning is from December to March in water at 0°C. One female may produce 3 million eggs, each 1·1 mm in diameter with a large oil globule. The eggs, transparent or yellow, lie separately on or near the bottom. They hatch in 3 – 5 weeks and the larvae are about 3 mm long. After one year they are 9 – 12 cm long.

2 **Silurus glanis** (Catfish, Wels) is a European fish found in the upper Rhine and in the rivers to the east of it that empty into the Black, Caspian, and Baltic Seas. It was introduced by the Duke of Bedford into the ponds at Woburn and has since spread to a limited extent. Usually found in lowland areas, it lives in deep, slow-running waters where the bottom is soft. It spends the day in deep water and emerges at dusk and in the early morning to hunt food. It is a predator and feeds upon fish, waterfowl and occasionally small mammals.

Spawning takes place, May to June, when the water is 18 – 21°C, after an upstream migration to shallow waters. Some 10 – 13 thousand light yellow eggs are deposited in a prepared nest. The male guards them for 8 – 14 days until they hatch. Young catfish are entirely black and tadpole-like in appearance. They remain in shallow water during the summer and feed on plankton. Growth is rapid before they move into deep water for the winter. After one year they are about 2·5 kg, and some about 30 cm long. At 6 – 7 years they are 100 cm. One fish of 81 kg was found to be 24 years of age. The record rod-caught specimen weighed 19·73 kg.

3 **Cobitis taenia** (Spined Loach) occurs in a restricted part of the south of England and throughout the rest of Europe. This small fish is a close relative of the stone loach (p. 100) which it resembles except that the head is flatter and the barbels on the snout are shorter. Below each eye is a groove in which a spine which when erect can prick sharply. Erect spines may penetrate the oesophagus of a predator and prove fatal.

It avoids still water preferring clear brooks with a bottom of sand or gravel where it lies concealed amongst stones or buried in sand with only the head and caudal fin exposed. It is a nocturnal fish which searches the bottom for its food — mainly worms, shrimps and larvae. It takes up mouthfuls of sand, retains any food it may contain, and expels the residue through the gill openings.

In the spawning season from April — June, the eggs are shed at random over the sand or amongst submerged roots. The young hatch in 6 – 10 days at 15°C. The adult may reach a length of 13·5 cm but is usually 11·5 cm.

1 BURBOT
2 CATFISH
3 SPINED LOACH

93

TENCH AND PERCH

The fish on this page live in still or very slow-running water of lakes, ponds, canals or rivers. The tench (4) belongs to the order Cypriniformes (p. 92) and to the family Cyprinidae, which includes the majority of European freshwater fish, characterized by the single dorsal fin, usually without spines. Large pharyngeal teeth, set behind the gill arches and under the gill cover, grind food against a horny pad at the base of the skull.

The other fish here (1 – 3) belong to the suborder Percoidei (p. 58) and the family Percidae and are found in inland waters. They have a broad dorsal fin, usually separate and composed entirely of spinous, hard rays, a second fin of soft rays, and a large anal fin below.

1 **Stizostedion (Lucioperca) lucioperca** (Pike-perch, Zander). First introduced in 1878 by the Duke of Bedford at Woburn, it has not spread far from these waters and appears to breed here only in favourable years. Spawning occurs in spring; one female may produce up to 300,000 eggs, each about 1·5 mm in diameter. Eggs are shed amongst weeds and the male guards them until they hatch in about 15 days at 12°C. The adult is usually 46 – 66 cm long and some 3 kg in weight, but the record rod-caught specimen weighed 6·945 kg.

2 **Perca fluviatilis** (Perch) has a general distribution throughout the British Isles, except for the north highlands of Scotland. In spring perch enter shallow water in shoals, segregated according to size. They swim near reed beds, where they spawn from April to July in water 2 – 7 m depth. The eggs, 2 – 2·5 mm in diameter with an oil globule, are laid in long strings, as flat bands or sometimes as a meshwork, and attached to weeds. They hatch in 10 – 18 days at 12°C and the larvae are 5 – 6 mm long and 1 mgm in weight. They can be found in large numbers in shallow, sheltered areas of water. In September, when they are 4 – 6 cm and 1 g, they migrate with perch of all sizes to deeper water for the winter which they spend in a sluggish condition. Until the larvae are 3 – 4 weeks of age, they feed on zooplankton. To this is added invertebrate bottom fauna until the larvae are 18 cm. There may be a short period of cannabalism before they reach 13 cm length. Larger perch eat small fish. Most males are sexually mature at 2 years, most females at 3 years, and all by 5 years when 16 – 17 cm long. A female of 13 years was 37·5 cm long and 1 kg in weight. The heaviest perch on record weighed 4·5 kg while the record rod-caught specimen weighed 2·154 kg.

3 **Gymnocephalus (Acerina) cernua** (Ruffe, Pope) has a limited distribution in the midlands, the south of England, and a small area of east Wales. In March or April they migrate from deep, still water, where winter has been spent, to shallow waters with reeds and sedges. Spawning takes place in April and May and the eggs are attached to the submerged plants. One female may produce 205,000 eggs, each 0·5 – 1 mm in diameter. They hatch in 12 days at 10°C and the young are 3 mm long. They feed at first on planktonic crustacea dnd later on aquatic insect larvae, particularly chironomids. After 2 years of rapid growth they are 9 cm long and sexually mature. Maximum length is 13 cm. The record rod-caught specimen weighed 113 g. Males may live for 3 years and females 4 years.

4 **Tinca tinca** (Tench) has a general distribution as far north as Loch Lomond and has been introduced into Ireland where it has a restricted distribution. Both males and females are mature sexually at 3 years of age. Mature males always have very much larger spoon-shaped pelvic fins than females. After migrating into shallow water, they spawn in June and July, sometimes also in August. Small groups of fish swim restlessly in the shallow water and then a female, often accompanied by two males, will break away and move shorewards. Spawning activity occurs amongst dense, submerged weeds, in which the eggs are widely dispersed. A series of separate bursts of spawning may occur with different males.

One female may produce 275,000 eggs, each 1·3 – 1·4 mm in diameter. They hatch in 5 – 7 days and the larvae, 4·0 – 5·5 mm in length, have pigmented eyes and a dense black stripe extending from the eye backwards along the gut. They hang from the vegetation with which they merge well. The yolk is utilized in about 10 days when the post-larva measures 6·0 – 6·5 mm. At 2 years they are about 7 cm long, at 4 years 15 cm, reaching a maximum of about 50 cm. A fish of 12 years was 50·1 cm and 2·495 kg; one of 7 years was 50·3 cm. The record rod-caught fish weighed 4·11 kg. There is considerable difference in the size of tench living in productive waters compared with those living in poor waters. Tench up to 14 cm feed mainly on *Cladocera*, copepods, and aquatic insect larvae. Larger ones take molluscs, caddis and chironomid larvae, Ephemeroptera nymphs, and *Asellus*. They are almost entirely carnivorous and many of the organisms eaten are inactive. They search over muddy bottoms for food at dawn and dusk guided probably by smell and by their rudimentary barbels.

1 PIKE-PERCH

2 PERCH

3 RUFFE

4 TENCH

CYPRINIDS OF SLOW-MOVING WATER

All the fish on this page belong to the super-order Ostariophysi (p. 92) and the family Cyprinidae (p. 94). They live in slow-moving rivers or lakes, generally on the bottom except the roach (2) which is a mid-water or surface fish. Just before spawning they move to shallow waters near banks as the eggs when shed adhere to weeds growing at the margin. Hybridization can occur between the rudd, the bleak, the bream, the silver bream, and the roach. Most hybrids are sterile but those of roach and bream, and of roach and rudd may be fertile.

1 **Scardinius erythrophthalmus** (Rudd) is common and widely distributed in Ireland and in the midlands and south of England. The majority mature sexually at 4 – 5 years of age, usually when 14·2 – 16·2 cm in length and 100 g or more in weight. Spawning takes place from April to June at which time the male becomes brilliantly coloured. The eggs are 1 – 4 mm in diameter. The young, at hatching, are about 4·5 mm long and hang from the vegetation. The average size of a fish of 3 years is 10·6 cm long and 42 g in weight and one of 9 years is about 20 cm long. Their growth is variable and in good water they may be much larger. The record weight for a rod-caught specimen is 2·041 kg. The young feed upon crustacea, and later chiefly upon plants, and aerial and aquatic insects. The rudd is an active carnivore and browsing herbivore. During the winter months it probably fasts.

2 **Rutilus rutilus** (Roach) is very common and has a wide distribution with the exception of Devon, Cornwall, west Wales, and north of Loch Lomond. It is present in Ireland in the river systems of Erne, Fairywater, and Cork Blackwater.
A few males are sexually mature at 1 year, but the majority at 3 years. Some females are mature when 2 years but most are 3 or 4 years. The short spawning season lasts only a week at the end of May and the beginning of June. The eggs, 1 mm in diameter, hatch in 10 – 14 days. The young are not very active until the yolk has been utilized, in 8 – 10 days, after which they live in dense shoals amongst the weeds.
Growth varies enormously even in fish of the same age from the same locality. The average length and weight, however, at 3 years is 11 cm and 28 g, and at 6 years 15 cm and 56 g. A fish may live for 12 years and would then be about 26 cm long. The record rod-caught specimen weighed 1·757 kg.
It is an omnivorous fish but generally takes more vegetable than animal food. Young ones eat planktonic crustacea and diatoms. There is a change from this diet to one consisting largely of insects, vascular plants, and algae about the time of sexual maturation. Molluscs are eaten only by older fish. There is a period of fasting during the colder months.

3 **Abramis brama** (Bream) has a wide distribution throughout the British Isles except in Dorset, Somerset, Devon and Cornwall, west Wales and north of Loch Lomond. The time of the onset of sexual maturation varies widely from 4 to 8 years of age. The youngest spawning fish on record were males and females 4 years of age, and 27 – 29 cm in length; the smallest were only 18 cm long and 5 years old. The spawning period begins in May and is complete before June. Shedding of the eggs is accompanied by much activity and leaping. One female may produce 200,000 – 300,000 eggs, each 1·5 mm in diameter, which hatch in 1 – 3 weeks and the young may be found with young silver bream in the shallows feeding upon diatoms and planktonic crustacea, particularly water-fleas. As they grow larger, insects, vascular plants, algae and molluscs are also eaten but form only a small part of the diet. Only fish 18 cm or more in length eat molluscs. In winter they fast for 3 – 4 months. Growth is very variable; the average size of a fish of 3 years is 16 cm and 70 g; one of 5 years, 20 cm and 154 g; and one of 8 years 30 cm and 500 g. They can attain a length of 41 cm at 12 years. The record rod-caught specimen weighed 5·840 kg.

4 **Blicca bjoerkna** (Silver or White Bream) has a limited distribution and is found in eastward flowing rivers from Yorkshire to Suffolk. In Norfolk it seems to be less common today than 100 years ago. Some are sexually mature and breed when 4 years of age, and all by 5 years. In the spawning season in May and June, shoals assemble in the shallows, and the fish are very active and will leap out of the water and splash at the surface. A female may produce 100,000 eggs, each 2 mm in diameter.
The young feed mainly on water-fleas till sexual maturation when they change from crustacea to insects and snails, thus competing with the bream for food. As the silver bream is a smaller species, it may be crowded out. It is relatively short-lived, the majority surviving for 5 years (13 cm) and only a few to 7 years (16·5 cm).

1 RUDD

2 ROACH

3 BREAM

4 SILVER BREAM

CYPRINIDS OF MODERATE-RUNNING WATER

The fish on this page belong to the super-order Ostariophysi (p. 92) and the family Cyprinidae (p. 94). They inhabit rivers and streams and breed in spring or early summer.

1 **Alburnus alburnus** (Bleak) can be found in Wales and in England south of the river Tees. A shoaling fish, it frequents the upper layers of water and can maintain a speed of 60 cm per second. It obtains food from the surface film, where it takes insect larvae including chironomids, *Simulium* and tipulids, as well as insects that have fallen on to the water. Algae and vascular plants are also eaten.

The male develops white tubercles at spawning time which lasts from April till July. The bleak gather in large shoals and move upstream to a place where the bottom is stony. The eggs, 1·5 mm in diameter, are deposited on plants or stones to which they adhere. They hatch in 14 – 21 days. Growth is fairly slow, a fish of 4 years being 9·2 cm, one of 6 years 13 cm, and a 7-year-old one 14 cm in length. They can reach 20 cm but the majority are smaller. The record rod-caught specimen weighed 111 g.

The scales of this fish were formerly used to make 'Essence d'Orient' for the production of artificial pearls. This trade is said to have begun in France in 1656.

2 **Leuciscus (Squalius) cephalus** (Chub) is found in England, except Cornwall and Devon, and in the eastern part of Wales. It lives in clean rivers and streams. The young are found in small shoals in the upper layers but older fish are usually solitary. Winter is spent at the bottom, preferably in deep water.

The onset of sexual maturation occurs in males of 3 – 4 years, and in females of 4 – 5 years. The males develop white tubercles during the spawning season which lasts from April to mid-July. One female may produce 50,000 to 100,000 eggs, each 1·5 mm in diameter, and these are shed over plants and gravel to which they adhere. Hatching occurs in 8 – 10 days and the young are about 7·5 mm long. Growth is slow initially but by 5 years a length of 25 cm is reached by males and 27 cm by females. Males of 10 years are 36 cm and females 40 cm. A length of 65 cm can be reached, and a weight of 3·5 kg. Young chub feed upon water plants, insects which have fallen on to the water, larvae of aquatic insects, worms, and molluscs. Larger ones feed mainly upon small fish, crayfish, and insects, as well as plant material.

3 **Barbus barbus** (Barbel) is found in Europe and has been introduced into some river systems in England, including the Bristol Avon, the Severn, the Medway, and the Thames. It lives in clear running water where the bottom is of sand or gravel. A bottom-living fish, it can sustain a swimming speed of 2·40 m per second and during the day will hold station against the current. At night it searches for food on the bottom including swan mussels, snails, worms, young fish, water plants, and detritus. Large shoals winter in quiet parts of rivers under vegetation. The males develop white tubercles in the spawning season which lasts from April to early July. They gather in large shoals and migrate upstream to places where the bottom is of coarse gravel or stones. The eggs, 2 mm in diameter, adhere to stones and hatch in 10 – 15 days. The eggs, or hard roes, are poisonous especially just before the breeding season.

The young grow fairly rapidly and at 1 year weigh 14 g, at 2 years 100 – 130 g, and at 3 years 550 – 600 g. A length of 100 cm and a weight of 6 kg can be attained, although they are generally smaller than this. The record rod-caught specimen weighed 6·237 kg.

4 **Gobio gobio** (Gudgeon) is found throughout England except the Lake District and Cornwall; and in east Wales; and is common locally throughout Ireland. It lives in shallow waters over gravel. A shoaling fish, it also exhibits territorial behaviour, returning to its 'home' if artificially transplanted. During storms and floods it shelters behind vegetation to avoid being swept away.

The males reach sexual maturity when 3 – 4 years old and 10 cm long, and the females when 2 – 3 years and 8 cm. The male develops white tubercles in the breeding season which last from April to July. At this time they frequent the beds of swift flowing rivers or the shallow margins of lakes. Groups of males and females may be seen to rush shorewards while shedding milt and spawn. The eggs, 2 mm in diameter, are sticky and adhere to plants or stones. They hatch in 14 – 28 days. Growth is fairly rapid in the first 2 – 3 years but subsequently slows. A fish of 3 years is 10 cm long and 13·6 g in weight, while one of 6 years is 13 cm and 30 g. A length of 15 cm can be attained. The males are larger, mature later, and die earlier, at 7 years, than females who usually live for 8 years. They feed chiefly on chironomid larvae, but also on other insect larvae including Ephemeroptera, Caddis, *Simulium* and Trichoptera, as well as small crustacea, *Asellus* and *Gammarus*. Plant material is also eaten and they feed throughout the year.

1 BLEAK   2 CHUB
3 BARBEL   4 GUDGEON

FISH OF SLOW-MOVING WATER

The bullhead (3), order Scorpaeniformes (p. 40), is the only freshwater member in the British Isles of the family Cottidae (p. 66). Of the two fish belonging to the order Cypriniformes (p. 92), the stone loach (2) is a member of the family Cobitidae (p. 92) and the minnow (1) of the family Cyprinidae (p. 94). They live in rivers or lakes.

1 **Phoxinus phoxinus** (Minnow) is found throughout most of Europe and the British Isles except for the northern highlands of Scotland. In Ireland it is common locally, its distribution expanding. It lives in streams, some lakes, and occasionally ponds. It is active from April to December and often swims in shoals of about 100 fish of all sizes, near the surface, feeding upon insects and their larvae, algae and diatoms. It winters in deeper water and eats less food then. Sexual maturity is reached at about 4 cm long, which a few reach at 1 year, the majority at 2 years. The male develops breeding colours in mid-March. His back is green-black, and the sides emerald green with vertical dark stripes (1B). The ventral surface is an intense scarlet, the head bronze-green, the throat matt black and the jaws have a touch of scarlet. The female's breeding colours are similar but less vivid.

The breeding season lasts from May to July with a peak in the first two weeks of May. Before spawning they migrate to shallow, running water where the bottom is gravel. Here they congregate in large numbers and are in a constant state of activity. Masses of 10 – 20 eggs, each 1·3 – 1·8 mm in diameter, can be found attached to the undersides of stones. A female of 2 years, 4·2 – 5 cm long, may produce 105 – 200 eggs; one of 3 years, 6·1 – 7·0 cm, 180 – 330 eggs. In 4 – 5 days the eggs hatch into fry 4·2 – 5 mm long. Young fry may be found swimming in shoals in quiet backwaters. Growth is slow; by December they are 23 – 24 mm long; after 1 year fish are 3·0 cm long and 0·3 g, and at 3 years 7·0 cm and 3·5 g. Most live for 3 years, a few survive for 4. It is one of the smallest of our freshwater fishes. The largest specimen recorded was 8·2 cm long and weighed 6·6 g.

2 **Noemacheilus barbatulus** (Stone Loach) is found throughout the British Isles except, possibly, the extreme north of Scotland, in still or flowing water. Though a poor swimmer, it can move swiftly along the bottom over short distances but spends much time dormant beneath a stone, becoming active at night. Bottom-living invertebrates are its chief food particularly chironomid larvae although aquatic larvae or nymphal stages of several insect groups as well as some adult crustacea are eaten. There is a reduction in food intake during winter.

They reach sexual maturity when 5·5 cm long which may be in the first year of life. The breeding season lasts April to July with the peak in May. They move into running water before spawning and eggs, 1·0 mm

in diameter, are often found scattered over the bottom. A female may have 5000 – 6000 ova; only about half are ripe, the others remain immature and are not shed. The newly hatched fry are 3·0 mm long, with black eyes and some pigment along the back internally below the vertebral column. The yolk is absorbed by the time the fry are 4 – 6·5 mm and the pectoral fins nearly 1 mm long. Growth is rapid and at 5 weeks they are 15 mm long, the barbels are present and they resemble the adult. At 1 year they are usually 4·7 cm long, at 2 years 8·0 cm, and at 5 years 11·4 cm after which little growth occurs, a fish of 6 years being 11·5 cm.

3 **Cottus gobio** (Bullhead, Miller's Thumb) is common only in England and Wales, rare in Scotland, and absent from Ireland. It lives in lakes and rivers, at depths of 6 m, where the bottom is of fine sand or gravel with small stones. A solitary, non-migratory, inactive fish by day, it lies in an S-shaped depression beneath a stone which it seldom leaves and to which it always returns, showing a strong homing instinct. In darkness it will search the bottom for food, stalking slow-moving animals which it swallows whole. Invertebrates form its chief food particularly larvae of the insect groups Plecoptera, Ephemeroptera and Trichoptera.

This fish may be mature and breed when only one year old, 4·5 cm long and about 1 g in weight. The male has the larger and broader head. When in nuptial dress, he is dark, and the anterior portion of the dorsal fin has a white or yellow edge. The breeding season is from April to mid-May. Before spawning the male enlarges the space beneath the stone and waits there until a female enters. The eggs may not be laid for 20 – 30 hours after pairing. This time the female spends upside-down, moving very little. A large and old female may lay up to 250 eggs, 1·6 – 2·0 mm in diameter, in a clump. More than one female may lay eggs in the same nest and even at the same time. The male remains to care for the eggs, fanning them with his pectoral fins to aerate them and keep them free from fungal growth. They hatch in 4 weeks at 10°C and the fry are about 6·5 mm long. By the time they reach 9 mm, the yolk is absorbed. The young disperse soon after leaving the nest and are only found singly. Much time is spent lying on the bottom during the first 4 weeks. At 2 years they are 5·5 cm and 1·5 g; at 4 years 6·5 cm and 4·0 g. In the Lake District these fish can reach 8·1 cm and 6·4 g, but in hard water they may reach 11·2 cm and 27·1 g.

1 MINNOWS (from above, and in section of water)   1A MINNOW. gravid female
        1B MINNOW. male, in breeding colours
2 STONE LOACH   3 BULLHEAD

FISH OF RAPID-RUNNING WATER

The dace (3) belongs to the super-order Ostariophysi (p. 92) and the family Cyprinidae (p. 94). The others (1, 2, 4) belong to the order Salmoniformes (p. 88). The rainbow trout (1) and the speckled trout (2) are members of the family Salmonidae which have a fleshy pointed flap above the base of the pelvic fin. The grayling (4) belongs to the family Thymallidae, characterized by the long-rayed dorsal fin the last ray being as long as those in the middle of the fin.

1 **Salmo gairdneri** (Rainbow Trout) is indigenous to the River Sacramento and its tributaries in North America. It has been introduced into England and Ireland but breeds successfully only in certain waters, the northernmost being the River Wye in Derbyshire. If young are reared in hatcheries and waters restocked as necessary, populations can be maintained but only in closed waters. Spawning occurs from November to May; male and female both cut the bed, or nest (p. 104), in which some 500 – 3,000 eggs are concealed after fertilization.

Young rainbow trout grow more rapidly than brown trout (p. 108), but they survive for only 4 – 5 years. The record for a rod-caught specimen is 3·855 kg, and for a 'cultivated' fish 4·543 kg. It may be about 50 cm long.

2 **Salvelinus fontinalis** (Speckled Trout) is indigenous to rivers of North America that drain into the Atlantic. It was brought to Europe in 1884 and later to the British Isles where it was introduced into lakes in Scotland. It has not thrived and is rather uncommon, although it can be cultivated to stock suitable waters. This inhabitant of cold, clear running water of mountain streams where the oxygen content is high can reach a length of 40 cm.

3 **Leuciscus leuciscus** (Dace) is found throughout England, except in Cornwall. Introduced into Ireland in the River Blackwater, Cork, in 1889, it is now also present in the Fairywater and Erne river systems. A fish of fairly fast flowing waters, it can sustain a swimming speed of 1·80 m per second. Most dace reach sexual maturity in their second year, when 10 – 11 cm in length, a few perhaps earlier. In the breeding period the head of the male becomes sprinkled with tiny greyish-white tubercles that resemble hoar frost. The spawning season lasts from February to May, varying with the locality. At this time they move into shallow water with stones or gravel at the bottom. A female 29 cm long, 400 g in weight, may produce 20,500 eggs. When shed, the eggs, 2·0 – 2·5 mm in diameter, adhere either singly or in clumps of 2 – 5 pebbles at a depth of 25 – 40 cm. The egg

capsule is greyish-white, relatively opaque and finely pitted, the yolk is biscuit-coloured. The incubation period lasts about 30 days at 11°C. The larva, 8·5 mm long at hatching, is slender and transparent. Within 24 hours a swim bladder develops and the little fish then hovers in the water. Soon afterwards compact shoals are formed and swim around fairly constantly. In 5 – 6 days they reach a length of 9·2 mm and at this size, just before the yolk sac disappears, they begin to feed, browsing upon unicellular algae and later on small crustacea such as copepods and cladocera. Metamorphosis is completed by mid-June when they are about 13 weeks of age. Small dace feed upon insects, diatoms, and plant material while larger ones eat snails, insects, freshwater shrimps, and plants. It has been ascertained, by the presence of tapeworms, that dace also eat tubificids which are the intermediate hosts.

At 1 year of age a length of 6 cm is attained, at 2 years 10·7 cm and at 3 years 15 cm. Subsequently the annual increment in length is smaller. A fish of 5 years is 18·7 cm; 7 years, 22 cm; and 9 years, 25 cm.

4 **Thymallus thymallus** (Grayling) is found in the southeast and midlands of England.

Sexual maturity is reached towards the end of the third year of life and they spawn at the beginning of their fourth year. Spawning, lasting from March to May, occurs in shallow water, 50 cm deep, over gravel beds. One female may produce up to 8,000 eggs, each 3 – 4 mm in diameter. The eggs are buried and incubation lasts for 20 days at 10°C. At hatching the fry are 10 mm long.

The rate of growth depends on the type of water. A fish of 2 years may be 14 cm long and 100 g in weight; one of 4 years 24 cm and 300 g; and one of 6 years 43 cm in length. One fish of 8 years has been recorded.

A variety of organisms is eaten, including insect larvae such as chironomids, *Simulium*, *Baetis*, small crustacea including *Asellus* and *Gammarus*, and molluscs. As much as 94 per cent of its food is bottom-living.

1 RAINBOW TROUT   2 SPECKLED TROUT
3 DACE   4 GRAYLING

THE SALMON

Salmo salar (Salmon). This fish belongs to the order Salmoniformes (p. 88) and to the family Salmonidae (p. 102). The salmon begins life as an egg lying in about 25 cm of gravel in a fast-flowing river. Development takes 80 to 90 days, at 7°C, and the young fish emerge, 16 mm long, with the yolk sac still attached. They are called *alevins* for 3 or more weeks until the yolk has been absorbed when about 26 mm, and then *fry* or *fingerlings* until finger length.

From now until they are about 15 cm long they are called *parr*. They have brown backs with black spots extending down towards the lateral line where a few red spots are present. Along each side there is a series of 8 to 11 dark patches, 'parr-marks'. The underside is light grey. Parr are indiscriminately carnivorous taking such food as small crustacea, aquatic insects and larvae, leaf-eating insects and gastropods.

After 1 year, in the Hampshire Avon for example, or 2 or more years in a more northerly river, they prepare for migration to the sea. Parr-marks are hidden by the deposition of guanine, a white reflecting substance, on the scales. The tail lengthens and becomes deeply forked and other changes take place. The *smolt*, as it is now called, is about 15 cm long and feeds ravenously during its descent to the sea. Large numbers travel downstream together, passing the many obstacles in their path. Until recently the life of the salmon in the sea had remained a mystery, but the tagging of large numbers of smolts led to the discovery of their feeding grounds, which abound with small crustacea and fish. Many salmon from British rivers travel between 3200 and 4000 km to West Greenland; one travelled at an average speed of 52·8 km per day over a distance of 1274 km. Growth is rapid and some reach 30 – 45 cm length and 454 – 681 g weight in 6 months. They spend a year, often much longer, in the sea and then begin the homeward journey for spawning. They do not eat again until spawning is complete. Staging areas exist on this journey where they gather in shoals before dispersing to various rivers. The majority return to their natal river — but how they locate it is not yet clear. The time of their return and run up the river varies; spring-fish rivers, such as the Welsh Dee, have a big run of salmon in March and April; summer-fish rivers, for example the Cumberland Derwent, have an autumn run. Salmon entering the rivers are beautiful, sleek, silver creatures which delight the angler and the fishmonger. If they have remained in the sea for four years, they may weigh as much as 19 kg. The record rod-caught salmon (1922) weighed 29·028 kg.

During the passage upstream obstructions have to be overcome, often by spectacular leaps. One leap of 3·4 m up a perpendicular fall was recorded. For such a leap the fish much reach a vertical speed of 9 m per second on leaving the water, which must be three times as deep as the height of the leap — 10 m in this case.

By October or November the salmon reach their spawning grounds. Females, dark in colour, are swollen with developing eggs; males, heavy with milt, are reddish in colour with elongated jaws, the lower one upturned with the tip — called the 'hook' or 'kype' — fitting into a socket in the upper jaw. Spawning grounds must have clean gravel of suitable coarseness, water velocity of 30 to 45 cm per second at the surface of the gravel, and a temperature of 2° – 6°C. A dominant male will prod a female with his snout, push against her, and drive off other males. The female will move out of the pool and after exploration will cut a bed (1B); she turns on her side alternately bending and straightening her body so that the tail flaps vigorously and dislodges the gravel immediately below it. The larger stones fall quickly to form a small mound downstream while the smaller stones are carried away by the current. A cut constitutes a complete series of movements, perhaps as many as 12 at 3 – 4 per second, beginning and ending with the fish in its normal position on an even keel. Cutting continues for hours or days until a saucer-like depression 15 cm deep has been made, at the bottom of which the current is much reduced. The female presses her body into the nest, her anal fin between two large stones, and raises her head and the front of her body away from the gravel. She now opens her mouth and the male slips forward until his head is alongside hers. The male opens his mouth and, with fins erect and opercula expanded, he quivers violently. Milt is ejected into the bottom of the bed and at the same time the female releases the eggs. Immediately afterwards the female moves a short way upstream and cuts vigorously so that the eggs are covered deeply. This forms the start of a new bed where the cycle is repeated with the omission of the exploratory stages.

The eggs, 5 – 9 mm in diameter when extruded, are sticky and adhere to the gravel. They absorb water rapidly and become spherical. The number produced is considerable; females of 5·0 and 8·3 kg produced 7750 and 8470 respectively.

Once spawning is completed the fish, now called *kelts*, are exhausted. Their bodies are often lacerated and covered with fungus, and the majority fall prey to disease. Too weak to keep station or make headway against the current, they generally drop downstream tail first. Examination of the scales of 11,455 Wye salmon revealed that 7·7 per cent had spawned twice, and of these only 0·7 per cent were males. Those that do survive return to the rich feeding grounds of the sea, coming back to spawn in the river again. A very small number survive to spawn three times; one 13-year-old, probably the oldest Scottish salmon on record, is known to have spawned on four occasions.

A salmon river must fulfil a number of conditions. The water must be clean and well aerated, the current swift, and the obstacles passable. Man has sometimes destroyed these conditions — effluent from the increased population and the industrial revolution has caused pollution, and many rivers have been obstructed. In recent years passes or ladders have been built, some of which have viewing chambers, for example the one at Pitlochry that bypasses the hydroelectric power station. The salmon has many natural enemies and care must be exercised to ensure the survival of the 'King of Fish'.

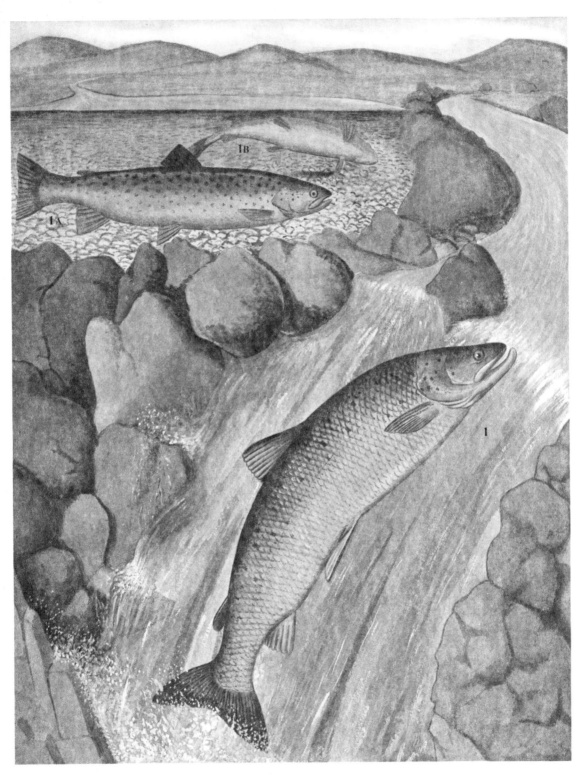

SALMON

1 Male.leaping while running up river   1A Female   1B Female.cutting a bed
(1A and 1B are shown in section through water)

THE SEA TROUT

Salmo trutta (Sea Trout). This fish belongs to the order Salmoniformes (p. 88) and to the family Salmonidae (p. 102). The species *Salmo trutta* has two forms, the sea trout and the brown trout (see p. 108). The sea trout is the form that migrates from fresh-water, where it hatches, to the sea and returns to freshwater to breed. The brown trout in contrast remains in freshwater, its migratory phase being restricted to movements upstream, or from a lake into a stream. Although the silvery sea trout and the brown trout differ in appearance and habits, they are similar morphologically. The difference between these two forms of *S. trutta* is best seen by examining under the microscope the growth rays in the scales. Those of the sea trout show a sudden expansion that reveals the migration they made to the sea with its richer supply of food.

Spawning begins in late September or early October and continues until February, with a peak in November. The fish usually wait in fairly deep water until they are ripe and conditions are suitable, when they swim rapidly upstream to spawn on pea-size gravel before the water level falls. The ova ripen in batches and are shed over several days. The bed is formed by the female (1A) making energetic flexions of the body and fanning with the tail. Stones and eggs, loosened by this activity, are swept downstream by the currents, and the male sheds milt over this growing mound from time to time.

The number of eggs produced varies considerably but averages 150 – 180 per 100 g of fish. The eggs hatch in about 81 days at 6°C. The yolk sac is present in the newly hatched alevin (1C) which remains in hollows between stones. The mouth is rudimentary but gradually develops so that after 3 – 8 weeks, by which time the yolk sac is absorbed, the alevin can feed and now ventures further afield. A few weeks later the 'finger' or 'parr' marks appear along each side; while they remain the young fish is called a *parr* (1B). The parr now disperse into water near the spawning grounds, or to a loch below, where the food supply is better.

The time spent in the river as a *smolt* varies between 2 and 6 years, but the majority migrate seawards at 2 – 3 years of age when 15 – 29 cm long; the sea trout smolt is thus larger and usually older than the migrating salmon smolt. It continues to feed on its migratory journey and there is also a change in its appearance. As the urge to swim seawards increases, the smolt becomes silvery to provide camouflage, for the colouration which provided protection in freshwater would be of no value in the sea. These silvery fish gather in large numbers in tidal waters, the same pools just above the tide being chosen year after year. Sometimes thousands of sea trout and salmon smolts may be seen packed together in these pools. After a few days in which to become accustomed to the change in salinity, they enter the tidal reaches or estuaries and move to and fro with the tide. Some may remain here for weeks or months, feeding, developing and growing, but others travel long distances. One marked fish, 280 g in weight, when recaptured 5 days later had travelled at least 3 km; another of the same size was recovered 10 months later when it weighed 700 g and had travelled at least 265 km.

Their prey includes sand eels, elvers, herring fry, sprats, and the fry of pollack, coalfish and other gadoids. If small fish are not available, then various invertebrates are eaten including annelids and small crustacea such as shrimps and prawns.

The smolts may re-enter freshwater in the same year, the majority in late summer, and now they are called *whitling*. Others, however, spend 1 – 3 years in the sea before returning to freshwater, some in March or April but the majority in June or later. Once in the river, they travel mainly at night, moving rapidly upstream. Most of these fish remain in pools up river or in lochs. The majority do not feed although some will eat worms and insect larvae. Their condition usually deteriorates with the time spent in freshwater; it may be said that they feed in summer and fast in winter.

Spring sees the return to the sea. Some 4 – 50 per cent of the whitling will have spawned, the number depending mainly on the river. The fish is called an *adult sea trout* from the second summer after migration, although some will have produced fertile ova or milt before this stage. The earliest age of sexual maturity is the third winter after hatching, and the latest on record is the eighth winter. Many sea trout survive to spawn again in successive or alternate years, mortality being much lower than in the salmon.

The oldest fish on record was 19 years; it weighed 5·67 kg, and came from a river where growth was slow. The oldest fish from a river where growth was fast was 14 years. The majority live for 10 – 12 years.

SEA TROUT

1 Male

1B Parr

1A Female

1C Alevins

THE BROWN TROUT

Salmo trutta (Brown Trout). This fish belongs to the order Salmoniformes (p. 88) and to the family Salmonidae (p. 102). The species *S. trutta* has two forms, the sea trout (see p. 106) and the brown trout in which the migratory habit has diminished considerably. However, it remains to some extent since stream-dwelling brown trout will move upstream to spawn, and lake-dwellers move into one of the inflowing rivers.

The brown trout lives in rivers with a steep gradient and consequently a very rapid current with plenty of dissolved oxygen. This is called the trout zone and is usually occupied by fish of the salmon family, which cannot survive in poorly oxygenated water. Because the water is fast-flowing the temperature, even in summer, remains cool. The lethal temperature for an adult trout is 26°C while the young cannot tolerate more than 23°C.

Lake-dwelling trout are generally found in deep lakes in mountainous regions. Colour and size vary in different waters. Those living in a mountain tarn never grow large, usually reaching a length of 18 cm and a weight of about 224 g. Brown trout 3 years of age can reach a length of 21 cm in Lake Windermere but in the River Avon in Wiltshire 31 cm.

A brown trout living in a mountain tarn usually has red spots on dark sides with a yellow underside, and the flesh is white. In contrast a lake-dweller is silver with black spots, and the flesh is pink. The striking colour variations are due to the brown trout's ability to imitate the background. Information about the environment is transmitted via the eyes to the brain. The colour changes are under neural and endocrine control; paling occurs rapidly as it is under nervous control, while darkening takes several hours, being controlled by the pituitary gland which produces a chemical messenger, a hormone. The brown trout swims by the side-to-side movement of its tail at an average speed of 4 – 5 m per second, but with bursts of up to 10 m per second when escaping from predators or pursuing prey. A carnivore, its eyes are important in the detection and location of prey. Aquatic invertebrates form the major portion of its diet, mostly insects and their larvae as well as molluscs and crustacea. Other fish may also be eaten. Its diet changes through the year as many of these organisms appear only seasonally.

Most trout mature when they are 3 or 4 years old. Maturation and spawning take place from October to February. The mature female has ovaries that almost fill the body cavity, forming as much as 20 per cent of the total body weight. When ready to spawn stream-living trout will move upstream, while lake-dwellers will shoal around the entrance of an inflowing stream. Such migrations are usually associated with a rise in the water level. The female will arrive at the spawning ground first and select a suitable site. She then begins to excavate the gravel by 'cutting', a procedure in which she turns on her side, alternately bending and straightening her body, to produce a vertical flapping movement of the tail. This activity causes the gravel to move downstream thus leaving a hollow. The male fights and chases other males from the territory, and then swims alongside the female and quivers while she continues until a nest about 7 cm deep has been cut. The female now remains with her pelvic region in the nest, her head held up and mouth open. Sperms and eggs are shed and fall between the loose gravel and stones of the nest. Egg-laying completed, the female moves upstream and again cuts, for 1 – 2 minutes; the gravel which is dislodged travels downstream to cover the eggs. Thus the start of a new nest is made, and several such nests will form a bed or redd. The spawning act itself lasts only a second, the eggs being fertilized as the sperms are shed as once in water the sperms survive less than one minute after extrusion. One female of 3 years was observed to lay 233, 220, and 141 eggs in three successive nests. A trout can spawn several times during its lifetime, but not necessarily each year. Lake-dwelling trout return to the same stream to spawn two or more times.

The eggs, 4 – 5·5 mm in diameter, settle amongst the gravel and develop, taking 27 days at 12°C or 77 days at 6°C. The newly hatched fish, 1·9 cm long with the yolk sac still attached, is called an *alevin*. For the first 2 or 3 days it remains buried in the gravel, and then begins to move about. The yolk sac is utilized in about two weeks when the fish is about 2·5 cm long. It is now called a *fry* or *fingerling* until it is a year old. In appearance it is streamlined and looks more like the adult. It begins to feed itself, swimming up towards small objects or prey such as midge larvae. From 1 to 2 years a trout is called a *yearling*, and from 2 to 3 years it is known as a *2-year-old fish*. At hatching the alevin has a naked skin. Scales begin to appear when the fry is 2·6 to 3·0 cm in length. The scales are of value in determining the age of many fish, including the trout. Bands or circuli formed on the scales differ in width when viewed under a microscope (p. vi), their size being largely dependent upon the abundance or scarcity of food. In winter the spaces between the circuli are narrow and the number of these regions indicates the winters through which the trout has lived. Occasionally trout survive for 10 or even 12 years but few live more than 8 years. The record rod-caught specimen weighed 8·221 kg.

Isaac Walton said in 1653 that the trout 'is a fish highly valued both in this and foreign nations'. Just over a century later a letter appeared in the Hanover Magazine from S. L. Jacob which gave detailed instructions for obtaining fertile eggs and rearing them in artificial conditions. There now exist a number of hatcheries where fertilized eggs and young are reared in large numbers for re-stocking existing trout streams or stocking new ones. The trout, greatly prized by anglers, has been introduced into streams and rivers in many parts of the world.

BROWN TROUT

1 Male   1A Female   1B swimming into the distance

FISH OF LARGE LAKES

All the fish on this page belong to the order Salmoniformes (p. 88), and 1 – 3 to the suborder Salmonoidei, (p. 102). The pike (4) belongs to the suborder Esocoidei, freshwater fish of North America and Eurasia. It is a predator and a member of the family Esocidae, characterized by the large mouth and teeth and wide, flattened snout.

1–2 **Coregonus** species (Whitefish) are found in some 15 localities in the British Isles. There are three main groups. The first includes the pollan of Lough Neagh, Lower Lough Erne, Upper Lough Erne, Lough Ree, and Lough Derg. The second consists of the vendace of Mill Loch, Castle Loch (now probably extinct), Bassenthwaite and Derwentwater. The remaining group consists of the powan of Loch Eck and Loch Lomond, the schelly of Red Tarn, Ullswater and Haweswater, and the gwyniad of Llyn Tegid. In each lake the population of *Coregonus* sp. has its own characteristic features. The common names have remained fairly constant but the systematic position of these populations is not yet clear.
Coregonus clupeoides (Powan) (1) spawns from late December to early January, on offshore banks of Loch Lomond over stones or gravel. The eggs, 3 mm in diameter, hatch in 60 – 70 days; the fry are 12 mm long. The average adult is 31 cm long and 230 g in weight. It may live for 10 years. Primarily a plankton feeder, it will also feed on small bottom-living animals.
Coregonus pollan (*albula*) (Pollan) (2) is found in Lough Neagh.

3 **Salvelinus** (**alpinus**) species (Char) of the British Isles are non-migratory lake colonies and have been so since the last glacial epoch some 12,000 – 14,000 years ago. The various forms have descended from one or a few migratory forms; they differ from *alpinus* and from each other and are mainly characterized by their proportions. Local names exist for the char found in different lakes. The natural history of *Salvelinus willughbii*, the Windermere char, was first described by Francis Willughby in 1686 in *De Historia Piscium*. In 1769 Thomas Penant in his *British Zoology* wrote of 'the remarkable circumstance of the different season of spawning in fish apparently the same'. It has been confirmed in recent years that in this lake there are two populations, one spawning in November on the lake shore in shallow water, the other spawning in spring in deep water of 18 – 24 m. A female spawning in autumn may produce 1220 ova, each 3·9 mm in diameter before fertilization. The eggs hatch in 64 – 80 days. A female spawning in spring may produce up to 1720 ova, each 4·3 mm before fertilization. The eggs hatch in 70 – 80 days.
Sexually mature char are at least 4 years and may be older. They frequently return to the same spawning ground. The female chooses a site where the bottom is preferably of walnut-sized stones, and then cuts a bed, or nest. The males fight for the female, and the dominant one circles and quivers against her while she continues to cut the nest until it is 4 – 6 cm deep. After about 7 hours the female crouches in the nest, the male alongside. The mouth is open wide in both and they develop great muscular tension before expelling almost simultaneously, in a few seconds, the ova and milt. Some 4 – 7 spawning acts follow in the same bed within 5 – 10 minutes; 25 – 250 eggs are shed in this time.
The char is a carnivore and feeds chiefly upon planktonic crustacea, only occasionally taking the larvae of aquatic insects.

4 **Esox lucius** (Pike) is found throughout the British Isles and Europe with the exception of Portugal and parts of Russia. The largest of our freshwater fish, it inhabits lakes and rivers, and sometimes ponds and streams. It lives amongst aquatic vegetation in shallow inshore waters down to 6·7 m. It is a solitary predator, chiefly of fish on which it begins to feed when only 40 mm long. The pike stalks its prey and when close enough, flexes its body, darts forward with great rapidity, seizes the fish crossways in its large mouth, turns it and swallows it whole, head first. There are many stories of large fish being taken; one pike of about 4 kg caught a salmon of similar size and 3 days elapsed before the entire body of the salmon disappeared from the pike's mouth. Although a voracious feeder, the pike can learn not to eat sticklebacks, particularly *Gasterosteus aculeatus* (p. 90), whose spines protect it.
The majority are sexually mature and breed at 2 years when males are 38 cm long and females 42 cm. Females of 2, 7, and 14 kg may produce 77,000, 200,000 and 270,000 ova respectively. The fish can be found on the breeding grounds in mid-March and spawning occurs, when the temperature is more than 6°C, in April and early May. Submerged plants must be present in order to evoke egg deposition in the female. The eggs, 2·6 – 2·9 mm in diameter, adhere to the vegetation. Hatching occurs in 14 – 21 days at 6 – 10°C, and the young are 7·5 – 8·2 mm long. They have a large yolk sac and a fastening organ on each side of the head for attachment to plants until the yolk has been utilized. The young subsequently feed upon small crustacea which live amongst the plants. Growth depends upon the type of water, the temperature and other factors. In Lake Windermere pike reach a length of 23 cm in their first year and a weight of 85 g. At 6 years females are 77 cm long and 4·73 kg in weight and males 67 cm and 2·66 kg. At 12 years females are 95 cm, and 9·22 kg; males are 79 cm and 4·53 kg. They can survive for 17 years in this lake. The largest one caught there was a female of 110 cm and 16 kg, but larger ones may be found in other waters.

1 POWAN 2 POLLAN
3 CHAR 4 PIKE

FISH OF SMALL LAKES OR ORNAMENTAL PONDS

All the fish illustrated on this page belong to the super-order Ostariophysi (p. 92) and the family Cyprinidae (p. 94). They all inhabit still or slow-moving water where the bottom is of mud.

1 **Carassius auratus** (Goldfish) is a native of eastern Asia and as early as the 13th century was bred commercially there. By the beginning of the 18th century it had reached Europe. Only the domesticated variety shows the gold or silver colouration due to the loss of black and brown pigment. They mature sexually in their second summer if conditions are good. The male develops tubercles at the beginning of the breeding season which lasts from April to August in the British Isles. They enter shallow bays where they spawn amongst water weeds to which the eggs adhere. The eggs are laid in batches, fewer being shed as the season progresses. They hatch in about 7 days at 20°C.
Adults can reach a length of 30 cm and a weight of about 900 g.

2 **Carassius carassius** (Crucian Carp) is found rarely in the south-east of England where it may have been introduced. It is a sluggish fish that ensures survival by burying itself in mud if the pond dries up. The breeding season is from April to June and the eggs, 1·4 – 1·7 mm in diameter, are shed amongst the weeds to which they adhere. Hatching occurs in 5 – 10 days and the fry at emergence are about 4 mm long. A length of 51 cm can be reached but 25 cm is more usual. The record rod-caught specimen weighed 1·991 kg.

3 **Cyprinus carpio** (Carp) is an Asiatic fish introduced into Europe in the Middle Ages. In 1496 Dame Julyan Bernes wrote in the *Boke of St. Alban's* 'there ben but fewe [carp] in England'. Now it is widely distributed in England, Wales, and southern Scotland. In Ireland it has been introduced locally but is not common. Sexual maturity is reached in males at 3 years of age and in females at 3 – 4 years. Tubercles develop on the head of the male at the beginning of the breeding period. They move into shallow water in May and June but spawn only at temperatures of 18°C or over. The numerous eggs, 0·9 – 1·2 mm in diameter, are laid in batches amongst reeds and water plants. The eggs hatch in 3 days at 20°C, and the fry are 4 – 5·5 mm long. In 13 weeks they reach a length of 30 mm and resemble the adults. To feed, carp suck up mud to a depth of 12 cm, digest the detritus and animals, and eject the inorganic material.
A large carp weighing about 20 kg was deposited in the aquarium of the Zoological Society of London in 1952

where it remained until it died in 1971 when over 30 years of age. They generally live for 12 – 15 years but some are said to survive for 40 years in captivity. The record rod-caught specimen weighed 19·957 kg.

4 **Rhodeus sericeus** (Bitterling), common in France, has recently been introduced into the British Isles and has a very restricted distribution. It feeds upon phytoplankton, crustacea, insect larvae, and tubificid worms.
This fish is best known for its symbiotic association with the swan or pond mussel, *Anodonta*, during the breeding period. The male develops tubercles on the head, the flanks become an iridescent steel blue and violet, and the throat and belly orange-red in preparation for the breeding season which lasts from April to June. The urogenital papilla of the female becomes protracted to form an ovipositor, often longer than the fish itself. This long, pink ovipositor is used by the female to deposit eggs, two at a time, into the mantle cavity of the mussel (4A). The male fertilizes the eggs by ejecting sperm near the inhalant siphon of the mussel. The eggs hatch and the larvae remain for 3 – 4 weeks, until the yolk is absorbed, in the mantle cavity of the mussel thus protected during that vulnerable period. The mussel also produces a crop of embryos, which often attach themselves to the bitterlings with attachment threads. Their early development takes place within cysts that they form on the skin of the fish, which in turn provides protection as well as a means of distribution for the young of the mussel.
The adult bitterling may reach a length of 9 cm.

5 **Leuciscus (Idus) idus** (Orfe, Ide), common in some parts of north-western Europe, has been introduced into England. It feeds on crustacea, insect larvae, water snails, worms, and other bottom-living invertebrates. Sexual maturity is reached in 3 – 4 years when about 15 cm in length. In the breeding season, in April and May, they congregate together and the male develops white tubercles. A female may produce up to 100,000 eggs which are shed on to the fine roots of weeds, or stones.

Micropterus salmoides (Large-mouthed Black Bass) has been introduced into England where it has become established locally in Dorset and Surrey. It belongs to the order Perciformes (p. 24) and the family Centrarchidae, freshwater fish of North America

1 GOLDFISH 2 CRUCIAN CARP 3 CARP
4 BITTERLING. Male 4A BITTERLING. Female, with long ovipositor
5 ORFE

· NEWTS

The class Amphibia is one of the smallest of the major vertebrate groups, with only some 2,000 living species. They were the first vertebrates to leave the water and invade the land, about 350 million years ago. The paired fins, with a bony skeleton, gave rise to the four limbs that enable them to move on land. The adults live on land except when they return to water during the breeding season. Newts, of the order Urodela, retain the long, fish-like form with a tail, but have two pairs of limbs.

1 **Triturus vulgaris** (Smooth Newt) widely distributed throughout the British Isles except in mountainous districts, is the only newt found in Ireland. Smooth newts, while terrestrial, live near ponds and eat mainly worms, slugs, snails and insects. When in water they take aquatic larvae, small crustacea, molluscs, and newts' and frogs' tadpoles. They detect their prey by sight and by smell, reacting rapidly. By the sudden expulsion of air from their lungs they produce a squeak, barely audible and of unknown purpose. Their skin is dry and roughish, giving a velvety appearance, and the markings are obscured. They hide in crevices and grass and emerge to search for food after dark. When the breeding season — March to early June — begins, sexually mature animals return to the pond and take on nuptial dress. The male develops a high, undulating crest from head to tail-tip. His back is dark olive green with black or dark green spots, his belly orange or yellow with dark green spots. The female's skin is finely granulated, dark olive above speckled with brown, while the belly is paler than the male's with smaller spots (1A). During courtship, the male after rapid movements eventually confronts the female with his tail bent double, its tip vibrating. He releases one spermatophore or more. This is a packet of several hundred spermatozoa enclosed in a whitish membrane, almost 2 mm long. The female then positions and lowers her body until the packet is pressed into the cloaca. The membrane breaks, releasing spermatozoa to be stored in the spermatheca of the female's reproductive tract for several days. The eggs then pass along the oviduct where they are fertilized and coated with mucus before being shed singly. Fertilization is thus internal, an advanced feature not common amongst the Amphibia. Up to 350 eggs may be shed over a period of weeks. Each, 3 mm in diameter, is attached to some object such as a leaf.

In 2 – 3 weeks the eggs hatch. The larva, about 6 mm long, has feathery external gills. The forelimbs emerge in 5 – 6 weeks and the hind limbs about 2 weeks later. Larvae remain in the pond until they undergo metamorphosis when 25 – 30 mm long; some reach this size at the end of their first summer, others not until the following May or June. These newts then leave the water and do not return until they are sexually mature in their third or fourth year. Maximum total length for males is 104 mm and weight 5·0 g; females, 98 mm and 5·5 g. It can live for at least 4 years.

2 **Triturus helveticus** (Palmate Newt) prefers high areas, up to 850 m, but may also be found in brackish water near the coast. It is the only newt in the mountains of Wales and western Scotland. It can be differentiated from the smooth newt during the breeding season; the female (2A) is a richer green, and the male develops a prominent ridge along each side of the back and a smooth-edged dorsal crest. Sexual maturity is reached when 60 – 75 mm long. During courtship the male follows the female pressing his lips against her cloaca; he then rushes in front of her, bending the distal half of his tail towards her, and vibrating it rapidly. If receptive, the female will remain staring at the male, till she accepts the spermatophore. Up to 437 spherical, fawn-coloured eggs are laid individually on leaves (from April to June in Hampshire). In about 21 days larvae about 6 mm long hatch. At first they feed on plant material but later on insects and small worms, like the adults. Metamorphosis begins when 25 – 35 mm long, and takes 2 – 7 days; the forelimbs appear, the gills become reduced, the skin becomes coarser, and other changes occur. It is the smallest of the European newts: the maximum size of males is a total length of 83 mm and weight of 3·5 g; females, 90 mm and 5·5 g.

3 **Triturus cristatus** (Warty or Great Crested Newt) is widely distributed in England and parts of Scotland and Wales. When seized, this newt exudes a noxious substance offensive to most predators from glands beneath the warty skin. The most aquatic of our newts, it prefers fairly deep ponds in clay or chalk, but it hibernates on land. It feeds upon small molluscs, insects, and aquatic larvae of insects.

Sexual maturity is reached in 2 – 4 years when 100 – 110 mm long. They return to the water in March, when the male develops a high, denticulated crest along the back and a separate one along the tail. In courtship the male keeps close to the female (3A), then obstructs her path while keeping his tail moving, and then deposits the spermatophore. The female moves over it and presses her cloaca upon it. Inside the female, the ova are fertilized by sperms released from the spermatophore. The eggs are laid one at a time, often on a leaf. A female produces 200 – 300 eggs, 4 mm long and 3 mm in diameter, that hatch in 3 – 4 weeks. Larvae (3B) from eggs laid in April may be ready to leave the water after metamorphosis when 50 – 60 mm long at the end of August (3C). This is the largest of the European newts; maximum length for males is 145 mm, females 162 mm; and maximum weights 10·6 g and 9·4 g respectively. One male of nearly 13 years has been recorded.

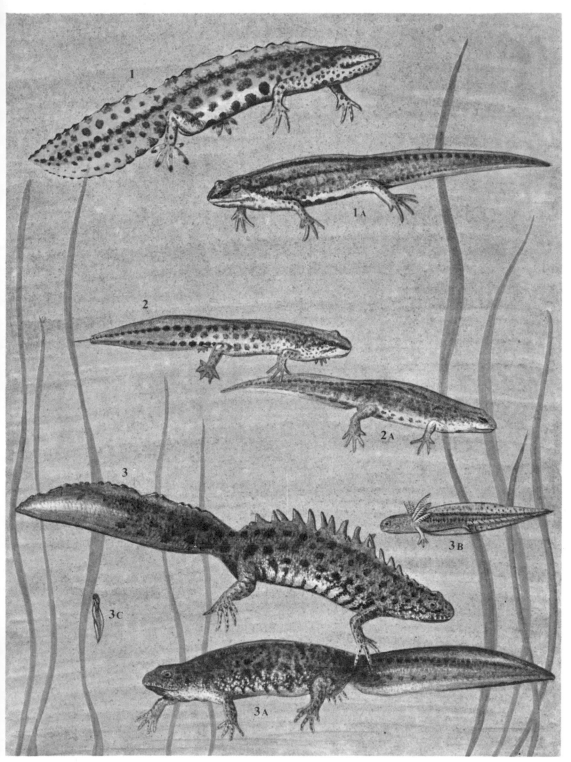

1 SMOOTH NEWT. Male
2 PALMATE NEWT. Male
3 WARTY NEWT. Male   3A WARTY NEWT. Female   3B larva

1A SMOOTH NEWT. Female
2A PALMATE NEWT. Female
3C recently hatched larva

115

FROGS

Frogs, class Amphibia (p. 114) of the order Anura, have a short backbone, no tail, and long hind limbs — all features adapted for locomotion by jumping — and webbed feet. The family Ranidae, including all on this page, usually live in damp places and have a moist skin. They can change colour to merge with the environment, by the expansion of pigment-filled cells.

1 **Rana temporaria** (Common Frog) is the most widely distributed of our Amphibia and there are few places where it is not found. In Ireland it was probably introduced in Norman times. It is less abundant recently, probably due to the loss of suitable wet-land habitats and ponds, increased human population density, and road mortality. During the breeding season it lives in shallow ponds and ditches, but spends much time on land during the remainder of the year.

It begins life in shallow water in early spring as an egg, 2 – 3 mm in diameter, within a mucilaginous envelope, 8 – 10 mm in diameter, amongst many others (1A). When first laid the eggs sink but as water is absorbed they rise to the surface. Exposed to sunlight, the surrounding jelly keeps the egg slightly warmer than the water it floats on. Development takes 14 – 21 days, depending upon temperature. To free itself, the tadpole digests the membrane with the secretion from a special gland. Behind this gland, on the ventral surface of the head, are the adhesive organs which secrete mucus that the tadpole (1B) uses as a means of attachment. Nourishment is provided by the yolk until the mouth forms in a few days and it can feed.

When the tadpole is about 5 weeks old, the hind limbs appear; by the seventh week joints have developed and also the forelimbs but they are not visible until the twelfth week, when metamorphosis begins. The external gills atrophy, the tail shortens, the eyes enlarge and develop lids, the tongue grows, and many other changes occur. Metamorphosis takes about 10 weeks, ending in May or June. The miniature adults, 12 – 15 mm from mouth to vent, spend much time out of the water sitting on small rocks. In October or November, when perhaps only 20 mm long, they begin to hibernate, mostly in holes or beneath stones away from the water. The adults, however, often bury themselves in mud at the bottom of ponds and ditches. Hibernation may not be continuous; they will feed if prey is available.

By the following autumn they are about 40 mm long. Sexual maturity is reached at 3 years and breeding begins soon after hibernation. The males arrive at the breeding pond first. Now and during mating they croak, usually in chorus, probably to guide the females to the pond. They already exhibit anatomical and behavioural changes which result from an increase in concentration of the male sex hormone in the blood plasma. The first digits of the forelimbs darken and develop a horny growth, the nuptial pad, which enables the male to hold the female. The male now reveals a strong urge to clasp the female — or lacking a female another male, a fish, sticks, or even a man's fingers. He places himself on the female's back and wraps his forelimbs around her body just below the armpits; this position is called amplexus. The female carries him around for about 24 hours, and considerable force is required to separate such a pair. Fertilization is external. The male releases the spermatozoa as some 1000 – 2000 ova are laid by the female. Ovulation is completed within a few seconds, the male releases his grip, and the female usually leaves the pond. After the breeding season, the adults live amongst plants near the pond. They remain almost immobile but if danger looms they escape by a number of leaps. The hind limbs, with their long, webbed digits, are efficient at swimming in water and at hopping or leaping on land.

Adult frogs feed mainly on insects, snails, and slugs, but also eat woodlice, butterfly and moth caterpillars, and beetles. The prey is caught by the rapid ejection of the long forked tongue which is attached to the front of the mouth.

The male may reach 73 mm and the female 85 mm, but they are usually smaller. In the 19th century a male of 80 mm and two females of 93 and 95 mm were recorded from Scotland. In captivity one frog which had been reared from spawn survived 12 years; it was 80 mm and 49 g.

2 **Rana esculenta** (Edible Frog) was introduced into East Anglia in the early 19th century, and later into southern England where small colonies may perhaps still be found.

3 **Rana ridibunda** (Marsh Frog), the largest European frog, was introduced from Hungary in 1934 to Stone in Oxney, Kent, and has spread into Romney, Welland and Denge Marshes, where it may have ousted the common frog. It feeds upon prawns, shrimps, newts, tadpoles, small fish, insects, and insect larvae. It reaches sexual maturity in 2 years and spawns in May and June. Each male has his own territory. He utters two calls throughout the season but with diminishing intensity. The female has a low-pitched call. In England males may be 96 mm long and females 126 mm but in Hungary it may be 170 mm.

Rana dalmatina is found only on Jersey where it is the only frog.

1 COMMON FROG   1A Spawn   1B new-hatched tadpoles attached to leaves
1C few days old, external gills developed   1D 4 weeks, external gills lost
1E 6 weeks, hind limb bud visible   1F-G 8 weeks, hind limbs developing
1H 10 weeks, fore limbs developed   1I miniature frog, tail almost lost
2 EDIBLE FROG   3 MARSH FROG

TOADS

The toads here belong to the order Anura (p. 116), and the family Bufonidae. Like frogs. toads can change colour to resemble their surroundings, but they are less aquatic and found in drier habitats. Their warty skin, with its noxious secretions, is less moist than the frog's.

1 **Bufo bufo** (Common Toad) is widely distributed from the Orkneys to the Channel Isles, but not in mountainous areas or in Ireland.

Toads sometimes hibernate in colonies. After emerging from hibernation, generally in March or April (7°C), they travel for 1 – 10 days to the breeding site, often in groups; at one site 2000 males were found. The colony will return year after year to one pond, probably their natal pond, in preference to others nearby; one colony has been found to persist in migrating to a now non-existent pond.

The males, smaller than the females, go singly to the breeding site, croaking to guide the silent females to them. A male will mount and grasp any female who comes, and she carries him 'in amplexus' (*see* p. 116) to the pond. His short forelimbs with the nuptial pads extend only to her armpit (axilla). Spawning will occur if the water is 9°C, at depths of 15 – 450 cm. The toads swim about in amplexus until the female finds some weed; she will then stop, with feet and arms on the weed. (This may be repeated several times.) The female stretches her body and extends her hind limbs slowly, and the male responds by stretching his hind limbs. Then the eggs emerge slowly, arranged in fairly regular rows of three, embedded in a long string-like mucilaginous coat (1A). Each length of egg-string, 2 – 3 m long, is extruded in about 10 seconds, the male ejecting sperm as it emerges. The pair now remain quiescent for several minutes and then move about amongst the weed. Oviposition occurs at intervals of some 30 minutes and lasts for several hours. The total number of eggs can be 3000 – 4000. The females leave the pond after spawning. Each egg is 1·5 – 2 mm in diameter and almost entirely black. After 10 – 12 days the embryo is about 4 mm long. The mucilaginous coat is partially digested and the embryo hangs motionless, attached to the remnants by its adhesive organ. When 20 – 27 days old and 12 mm long, the tadpole swims freely (1A). Metamorphosis is compete in 65 – 108 days, depending upon temperature. The tadpole leaves the water when 10 – 15 mm, and leads a solitary life nearby under stones or leaves. By autumn, it is 20 mm long, and begins hibernation. By next autumn it is 40 mm. Toads feed mainly in the evening, on moving prey only, including beetles, ants, worms, snails, insects and their larvae. The long unforked tongue with its adhesive tip is flicked in and out in about 0·1 seconds.

The toad avoids predators mainly by its excellent camouflage. If frightened, it may inflate and so look bigger, at the same time extending the hind limbs, lowering the head, and raising the rear of the body off the ground. This manoeuvre deters predators such as weasels and hedgehogs. The secretion of glands in the warty skin is noxious to many animals; it causes irritation of the mouth, a slowing of the heart, and paralysis of the muscles. The female is sexually mature at 4 years, the male at 3 years. The male reaches full growth in 5 – 6 years, the female may take a little longer. The largest recorded were a male 74 mm, and a female 99 mm weighing 148 g. Those on Jersey are larger, one of 12 years being 120 mm. In the 18th century one of 36 years was recorded and more recently one of 20 years.

2 **Bufo calamita** (Natterjack) occurs in some places on the east coast from Yorkshire to Kent, the Solway Firth to North Wales, County Kerry, and parts of southern England.

Confined to coastal areas, it buries itself in sandy soil by excavating a tunnel with the strong forelimbs. It is a gregarious species and usually stays near the breeding site, although a whole colony may migrate to another area. It is mainly nocturnal. It is smaller than the common toad, with shorter limbs which do not permit it to hop; instead it has a running gait. Its diet is similar to the common toad's. It hibernates from late October to February. The long breeding season lasts from mid-April to the end of June. They spawn in shallow water and the egg-strings, about 2 m long, contain 3000 – 4000 eggs, each 1·5 mm in diameter with a mucilaginous coating 3 – 4 mm thick (2A). They hatch in 5 – 10 days and metamorphosis is complete in 6 – 8 weeks. On leaving the water, the young resemble the adults but are only 7 – 10 mm long from snout to vent. By autumn they are about 20 mm, and attain full growth in the fourth or fifth year. The largest recorded were a male 70 mm long, and a female of 80 mm.

Hyla arborea (European Tree-frog) was introduced in the Isle of Wight in 1906; there is also a colony in south Hampshire.

Hyla ewingii (Ewing's Frog) was introduced in west Cornwall in 1951 from Tasmania.

Bufo viridis (European Toad) was introduced in the Isle of Wight in 1958.

Alytes obstetricans (Midwife Toad) has flourished since 1900 in a garden in Bedford, where it was introduced. Colonies have since been introduced into south-west and mid-west Yorkshire. Among its interesting habits, the male carries the eggs wrapped around his legs during development.

1 COMMON TOAD
2 NATTERJACK TOAD

1A COMMON TOAD. spawn and tadpoles
2A NATTERJACK TOAD. spawn and tadpoles

TURTLES

Members of the class Reptilia are characteristically tetrapods (four-limbed animals). In some lizards and all snakes, however, the limbs are greatly reduced or lost altogether. Horny scales cover the body. Fertilization is internal and the eggs, encased in a shell (usually leathery), are laid on land.

Turtles belong to the order Chelonia; they have a short, broad body encased in a shell. This consists of a carapace above and a plastron below joined along each side but leaving front and rear apertures. They lack teeth and the horny jaws are beak-like. They are natives of tropical and sub-tropical regions where, after copulating in the sea, the female comes ashore to lay eggs in holes dug in the sand. The word 'turtle' is generally used in Britain only for marine chelonians, but in America it is often used for all chelonians, or at least all aquatic ones.

1 **Caretta caretta** (Loggerhead) is occasionally stranded along the south and west coasts of Ireland, and from Devon and Cornwall northwards to the Shetland Isles, as well as on the east coast of Scotland and southwards to Scarborough. Of the 40 – 50 recorded the majority have been young specimens, perhaps swept from coastal waters in the Caribbean by hurricanes.

Their breeding sites, more northerly than other turtles, are on the eastern coast of America south of North Carolina and on the Atlantic coast of Morocco. After excavating a pit 45 cm deep in the sand, the loggerhead lays 120 – 150 spherical, soft-shelled eggs (40 mm in diameter and 43 mm long), which all hatch at the same time after 31 – 65 days. The young, resembling miniature adults, try to climb from the pit, and dislodge sand which raises the floor level thus helping them to the surface. After leaving the shelter of the nest they are vulnerable to attack from many animals, both on their way to the sea and in it. Man also reduces their numbers as he harvests the eggs although he rarely eats the flesh. Adults of this large species may reach a total length of about 2 m, with a carapace of 1 m. They are carnivorous and usually eat crabs and molluscs.

2 **Dermochelys coriacea** (Leathery Turtle) normally inhabits tropical and subtropical seas. There are 42 records of it between 1756 and 1966, all around the British Isles. The largest one, 350 kg, was taken off Ireland. It can attain a total length of 270 – 300 cm, with a carapace of 180 cm. The largest of the living Chelonia, the order which includes both turtles and tortoises, it may be twice as long as a giant tortoise. The carapace has seven ridges running longitudinally along the back and it tapers to a tail-like process. The thick leathery skin covering the carapace gives this turtle its common name. Beneath the leathery skin is a shell formed from a mosaic of bony platelets.

There is a single notch in the centre of the upper jaw into which fits a tooth-like projection on the lower jaw. On either side of the notch the upper jaw is emarginated, giving the impression of two teeth. Although little is known of the feeding habits, this turtle appears to be omnivorous. It is certain that jellyfish and the Portuguese Man-o-War, *Physalia*, are eaten and perhaps the notched jaws help the turtle to grasp their thin soft slippery bodies. This turtle is a considerable traveller and is probably a periodic visitor to cooler waters. It has been recorded as far north as Nova Scotia in the western Atlantic, and to Scotland in the east. Perhaps these extensive journeys in open sea are possible because it can live on pelagic jelly-fish. When swimming at the surface, this turtle sometimes raises its head and neck out of water. It has been found to be accompanied by *Naucrates ductor* (p. 30) the pilot fish, while *Remora remora* (p. 42) has been found attached to its shell.

It breeds on a number of coasts, including those of Malaya, Surinam, Costa Rica and Florida; the breeding season varies with the locality. In Florida a female laid 80 almost spherical eggs, 50 – 54 mm in diameter, at a depth of 1 m some 45 m away from the water. Eggs are generally laid at night and the exhausted female can return to the sea under cover of darkness. Incubation lasts 58 – 65 days and at hatching the carapace is 58 – 60 mm long.

The adult is only rarely eaten — in some areas the flesh is considered poisonous at certain times. In contrast, the eggs are frequently taken by man, both as a source of protein and as an aphrodisiac.

Lepidochelys kempii (Kemp's Ridley). Of the 18 recorded around the British Isles, all have been fairly young specimens. One with a carapace 20 cm, weighing about 2 kg, was stranded in Ireland in 1968. It is a small species with carapace up to 68 cm and weight up to 45 kg.

It is an inhabitant of the western Atlantic from the Gulf of Mexico northwards to Massachusetts. Breeding grounds on the coast of Mexico were recently discovered. They breed in May and June and on one occasion some 40,000 were observed, gathered on a beach. The female, unlike other species, lays her eggs during the day. Usually 100 almost spherical eggs, each 38 mm in diameter, are laid in a hole 37 cm deep, and they hatch in about 60 days. Little is known of its feeding habits.

1 LOGGERHEAD TURTLE
2 LEATHERY TURTLE

LIZARDS

Lizards and snakes, of the order Squamata, have paired copulatory organs and fertilization is internal. Most of the lizards have four limbs, movable eyelids, an external opening to the ear, and a tongue which is sometimes forked — as in all the British species — but is often rounded or only notched at the tip. They feed mainly on insects. The lizards on this page, except the slow worm (1), belong to the family Lacertidae which are mostly small or moderate in size. The family Anguidae includes both four-legged and snake-like forms such as *Anguis* (1), in which the only remnants of limbs are inside the body.

1 **Anguis fragilis** (Slow Worm) is widely distributed in dry habitats in England, Wales, Scotland, and the Channel Isles. It hibernates in winter in holes or beneath stones.

During the breeding season, from the end of April to June, the males fight, seizing one another by the head or the front part of the body, and inflicting wounds. In mating the male seizes the female by head and neck, and twines his body around her so that the vents meet. They may remain in contact for several hours. The female retains the fertilized eggs in the oviduct for 3 months or more until the young are nearly ready to hatch. One large female gave birth to a litter of 19 live young, and one which failed to emerge from the membrane, 8 – 9 cm in length and 0·4 – 0·6 g in weight. Perhaps the largest number recorded in one female was 26 fertilized eggs. When 1 year old they are 15·2 – 18·0 cm in length and at 2 years are 21 – 23 cm.

The young at birth have a characteristic colouration (1B). As they grow so the colours fade, until in the male they disappear. These changes begin in the third year and differentiation is marked by the end of the fourth year. The male is said to be sexually mature when 3 years and the female a year later. Adults are usually 35 – 40 cm long although a large male of 48·9 cm was found in Hampshire. Under very favourable circumstances it has been known to survive for 54 years. It feeds only on moving prey, chiefly slugs and earthworms, but will also eat other small animals.

As its name implies, most of the actions of this animal are deliberate and slow, although if in danger it can move rapidly, with a gliding motion.

2 **Lacerta vivipara** (Viviparous or Common Lizard) is widely distributed in England, Scotland and Wales and also in Ireland where it is the only reptile. It lives in all types of habitat from sand dunes to mountainous areas.

During the breeding season, in April and May, large numbers sometimes live together, although the males may fight each other. When ready to mate a male will seize a female in his jaws, by the head or body, and manoeuvre until he holds her near the groin. Copulation may last 5 – 30 minutes. The fertilized eggs are retained within the oviducts for about 3 months, and the young hatch at or soon after laying. The average litter is 5 – 8, the maximum 11. The young are usually deposited in a cavity made by the mother in a damp place, and concealed. Some hours after birth the young run about and will feed on small spiders or green aphids. They are dark bronze-brown above, greyish-black below, with back and sides sometimes gold-speckled. At birth they are 3·7 – 4·7 cm long and by autumn, when they enter hibernation, the largest are 7·8 cm long. By the following summer it is possible to distinguish the sexes. The male has a swelling, due to the copulatory organs, at the base of the tail and may also show a more strongly patterned colouration. The male can attain a length of 17 cm, and the female 17·8 cm. One has been kept for 5 years in captivity. Its chief food is spiders, centipedes, insects and their larvae. It can move rapidly, in a series of dashes, making it difficult to catch. If a predator seizes its tail, then the lizard will sever it; this ability, known as autotomy, is due to weakness of special regions of the vertebrae. It is agile and can climb fairly well, and it may sometimes be seen basking with belly on the ground and limbs sprawled out.

3 **Lacerta agilis** (Sand Lizard) is restricted largely to dry heaths and sand dunes in Dorset, Hampshire, and the Isle of Wight but its numbers are being reduced by the destruction of these habitats. It feeds chiefly upon insects.

Females may reach a length of 18·5 cm, males 19·3 cm. The adult male is easily distinguished by the green colour on the underparts, most vivid in the breeding season. In May or June the female will lay 6 – 13 oval eggs, 8 mm in diameter, in a concealed hole which she has dug. The eggs hatch in 7 – 12 weeks and at birth the young, 5·6 – 6·3 cm long, resemble the adult in colouration.

Lacerta viridis (Green Lizard). This native of Jersey where it is fairly common has very bright colouration. It is a large species, up to 35 cm long.

Lacerta muralis (Wall Lizard) is found only on the north-east coast of Jersey.

1 SLOW WORM. Male 1A SLOW WORM. Female 1B SLOW WORM. new-born young
2 VIVIPAROUS LIZARD. Male   2A VIVIPAROUS LIZARD. Female
3 SAND LIZARD. Male   3A SAND LIZARD. Female

SNAKES

The order Squamata (p. 122) includes snakes which belong to the suborder Ophidia. The body is elongate and limbless. Some of the jaw bones are connected only by loose ligaments, and the jaws, owing to their loose articulation, can move very freely to open the mouth wide and swallow prey whole; the tongue is deeply forked and protrusible. The eye is covered by a transparent spectacle and there is no movable eyelid. The ear has no external opening. The great majority of snakes belong to the family Colubridae, which includes the grass snake (1) and the smooth snake (2). The adder (3) belongs to the family Viperidae. No snakes are found in Ireland.

1 **Natrix (Tropidonotus) natrix** (Grass or Ringed Snake) is not poisonous. It is found in woodlands, hedgerows and marshland, preferably in districts near water, in England and Wales. Frogs are its chief food but newts, toads, and sometimes small mammals are eaten; it will also swim in search of tadpoles, fish and lampreys. After hibernating from October to March or April, it emerges just before the breeding season. Males are sexually mature when about 50 cm long, females when 73 cm, at 3 – 4 years of age. Mating occurs in April and May. A fully grown female lays 30 – 40 eggs at the end of June or early in July in a chosen site where warmth is artificially generated, such as a heap of rotting leaves or a hayrick. Several females may use the same site; one such nest contained 250 eggs, and another 1200. Hatching occurs in 6 – 10 weeks and the young are 16 – 19 cm long. Adult males may be 81·5 cm long, and females 110 cm; larger ones, probably females, up to 172·5 cm have been recorded.

2 **Coronella austriaca** (Smooth Snake), non-poisonous, is rare, and restricted to Dorset, Hampshire, Surrey, and one record from the Isle of Wight. It inhabits heath and woodland with sandy soil. It feeds chiefly upon lizards, but slow worms, small snakes, field mice, voles and shrews are also eaten.
Little is known of its breeding habits in this country. One captured female produced 5 young, each encased in a semi-transparent membrane, at the end of September, but the young did not survive the winter. Males can reach 56·3 cm in length and females 65·1 cm.

3 **Vipera berus** (Adder, Viper), the most common British snake and the only poisonous one, is found in England, Scotland and Wales. It is easily recognized by the distinctive zigzag pattern along the back, but is difficult to detect on the dry heaths where it lives. Different habitats are occupied depending upon the season. It may hibernate in an area of dry elevated ground with thick vegetation, but spends the summer on lower ground. It moves by horizontal undulation, and journeys of up to 1500 m have been recorded between two habitats.
Males reach sexual maturity when 40 cm long, at 4 years. Eggs begin to develop in the female when she is about 47 cm long, and are fertilized in the following year. The female usually breeds in alternate years. Some weeks after leaving hibernation mating occurs, in April and early May. At this time the 'dance' of territorial rivalry may be seen between two males. Raising the front part of the body, and swaying their heads in contrary directions, they entwine each other. Eventually one snake disengages and leaves the territory pursued by the victor. The male will accompany the chosen female, crawling beside her and tapping her back with his jaw, and licking her body frequently. They occasionally dash forward, the front part of their bodies raised off the ground, while the male tries to get his tail beneath the female.
The adder is ovoviviparous. It is thought that the egg-chambers of the oviduct become highly vascularised and something resembling a placenta develops. During the gestation period of 4 – 4·5 months the female often basks in the sun. In August or September some 6 – 20 young are born. At birth the 12 – 20 cm embryo if still enclosed in its membrane, ruptures it by convulsive movements usually with a final thrust of the snout. In autumn, usually October, adders enter hibernation, singly or in numbers, in holes or cavities. This lasts about 180 days during which it neither feeds nor grows. Growth occurs mainly between March and October. After one year a length of about 26 cm is reached and after 2 years 35 cm. Females may attain a length of 74 cm and males 60·6 cm.
They feed on a variety of prey especially rodents, young birds, lizards and worms. The snake strikes its prey, injects venom, and then releases the prey which soon dies. The venom is discharged via the tube, formed by infolding, in the erected, backward curved fang at the front of the upper jaw. After a strike the mouth remains wide open, and the viperine fang returns to its recumbent position. The poison is haemorrhagic, acting on the small blood vessels; it causes considerable swelling of the part bitten, and is also slightly neurotoxic. The snake does not poison itself as the secretion has to enter the vascular system before it can act upon the organism.
In the past 50 years some 50 people have been bitten in Britain, of whom about 8 have died. Bites occur mainly in April and May when the snakes are moving from the hibernation area and mating. But at any time if an adder is trodden upon, it strikes in self-defence. The victim usually feels a sharp pain, which may be followed by oedema and vomiting. The severity of the symptoms is proportionate to the amount of toxin introduced and to the size of the victim. Therefore, children suffer most. Anyone who is bitten should seek medical advice immediately.

1 GRASS SNAKE
2 SMOOTH SNAKE
3 ADDER

HEDGEHOG AND MOLE

The class Mammalia is characterized by mammary glands, hair on the body, warm blood, and a relatively large brain. The two members of the order Insectivora shown here have survived with little change since the Eocene period. The mole (2) belongs to the family Talpidae and the hedgehog (1) to the family Erinaceidae.

1 **Erinaceus europaeus** (Hedgehog, Urchin) is widely distributed throughout the British Isles and is present on some of the islands. In Ireland it may have been introduced for food. It is common in most open country, on grassy heaths, cultivated land, and scrub, and on sand dunes wherever there is cover for nests and shelter; but it is rare in marshes, dense woodland, and areas of Wales and Scotland over 450 m. It abounds in parks and gardens in urban areas.

The spines which cover the dorsal surface are the most characteristic feature. The hair has become modified into sharp pointed spines, about 16 per square cm. Each is 2 – 3 cm long and 1 mm in diameter, with a knob-like base which is inserted into a thick layer of muscular tissue. The spines are erectile, and, combined with the hedgehog's ability to curl into a ball, they confer considerable protection. It is the least preyed upon of the small mammals. Man is its greatest enemy and the roads bear silent witness to the effect of the car on this species during its nocturnal wanderings.

The breeding season begins shortly after they emerge from hibernation. Both sexes are mature when 1 year of age. The males are fertile from early April until August, and the females are pregnant between May and October. Courtship is somewhat protracted. Facing each other, one advances while the other retreats and then the first retreats while the other advances. This is accompanied by considerable snorting. The male, in spite of the spines, mounts the female as do other quadrupeds. The male's penis is very large for an animal of this size. The gestation period lasts for 30 – 40 days. There is a peak in the birth rate in May and July and a smaller one in September.

The young, usually 5 to a litter, are born in a large breeding nest. Two litters may be produced in one breeding season. At birth the young are blind — their eyes open at about 14 days. The spines, arranged in a regular manner, are soft and white. They weigh 11 – 25 g and are suckled for 4 weeks. Growth is rapid: after 7 days they weigh 22 – 50 g; after 128 days 650 – 1125 g. At 3 weeks they begin to leave the nest but if they stray too far the mother retrieves them in her mouth. Finally they follow the mother in a string to go on a foraging expedition.

The head and body of the male is 188 – 263 mm, the female 179 – 257 mm, the tail 26 and 23 mm respectively. The male weighs 900 – 1200 g, the female 800 – 1025 g, varying seasonally. It may survive 7 years.

The hedgehog is active at night when it searches for food. Its diet includes insects, their pupae and larvae, slugs, beetles, earthworms, snails and vegetable matter. While foraging, many grunts and snorts are emitted;

these sounds may be the first indication of their presence in the garden. They can be encouraged to pay regular visits if bread, milk and a little meat are put out each evening. During the late autumn they feed well and masses of brown fat are laid down in the region of the neck and shoulders. These stores provide material to meet metabolic requirements.

2 **Talpa europaea** (Mole) is found in England, Wales and Scotland, and in the islands of Anglesey, Mull, Skye, Wight, Alderney, and Jersey. Most abundant in deciduous woodland and adjoining areas, it can also live in regions up to 1000 m.

Its presence in pastureland is revealed by 'molehills' (2B), the product of its activity when building tunnels 5 – 100 cm below ground. Other types of digging include shallow or surface tunnels where the displaced soil forms a ridge, and the open, rutting or 'trace d'amour' type where the surface is burrowed energetically and the soil is thrown aside to form a deep groove. In digging the mole uses each of its large forelimbs in turn, changing after a few strokes, to scrape the soil to the rear. At intervals it moves the accumulated soil in front of itself by sideway movements of the forelimbs.

Highly specialized for burrowing it has a massive pectoral girdle; the large forelimbs have broad digging claws which grow and are worn down continuously. It has no external ears, only rudimentary eyes, and smooth, fine, velvety fur. Being an active animal it needs much food, and eats mainly earthworms, insects, myriapods and molluscs. These it gathers in the tunnels, finding them there as trespassers or digging for them.

The majority of females bear only one litter per year, usually of 4 young. The breeding season lasts from February to June depending upon the region. The peak in birth rate is in April and early May, after a gestation period of about 4 weeks. The period of lactation also lasts about 4 weeks. At birth the young, 3·5 – 4 cm long excluding the tail, and 3·5 g in weight, are red and hairless. After 14 days the body appears black because of the development of pigmented hairs. At about 3 weeks the eyes are open, the fur well grown, and the young weigh about 60 g. Already the males are larger, weighing as much as 10 g more than the females. They do not reach sexual maturity in their first year. There is evidence that 10 – 15 per cent of females survive for 3 years. A length of 16 cm for the head and body can be reached but there is considerable variation in the weight of moles from different regions. Males from the south weighed 95 – 110 g whereas ones from north Wales weighed 80 – 95 g; females weighed 72 – 81 g and 62 – 64 g respectively.

1 HEDGEHOG with new-born young
2 MOLE   2A MOLE, new-born young   2B molehills

SHREWS

These animals belong to the order Insectivora (p. 126) and to the family Soricidae, all small forms, that includes the smallest known mammal *Suncus etruscus*, found in southern Europe, as well as the smallest British one (2). They are distinguished by the large incisor with a hook at its end and a ventrally projecting cusp at its base. These mainly nocturnal animals have small eyes, soft fur, and long snouts with sensitive vibrissae, used in the search for food.

1   **Sorex araneus** (Common Shrew) is found over most of England, Wales, and Scotland and their islands. It lives in most habitats up to 1000 m but prefers thick grass, woodland, scrub, hedgerows and banks. Extensive runs are made in the litter zone either by digging tunnels or using those of other species. Since the surface area of the shrew is large in proportion to its weight, it needs a lot of food — a time-consuming pursuit. Beetles, earthworms, snails and other invertebrates of the leaf litter, as well as plant material are eaten. Its hunting movements are swift and almost invisible, the sensitive vibrissae around the snout being the chief food-finding organs while soft, twittering sounds are emitted during the search.
Sexual maturity is reached at the end of the first winter or early in spring. They breed from April to October, producing several litters each of 5 – 8 young. The gestation period lasts for 13 – 19 days. The young, weighing less than 0·5 g, are born in nests of grass woven into a loose ball. They develop slowly; the eyes open in 18 – 21 days when they are about 7 g, a few days before weaning. The rate of growth varies throughout life but there is a final burst of growth at sexual maturity when the maximum size is attained. The head and body is 74 – 85 mm long, the tail 34 – 42 mm, and the weight 8·5 – 13·6 g. Two moults occur, one in the first autumn to replace the lighter coloured fur with a long, dark winter coat; and then a spring moult with shorter hair of the same colour. The adults die when just over 1 year of age.

2   **Sorex minutus** (Pigmy Shrew), the smallest of our mammals, and the only shrew in Ireland, is widespread elsewhere but absent from the Scillies, Shetland, and Channel Isles. More common in open habitats, it will live anywhere from mountain regions up to 1500 m, to sandy coastal areas. Its food is probably similar to the common shrew's.
This solitary animal has a territory range of about 30 m. It is only aggressive if contact is made with other shrews outside the breeding season. More active by day than by night, it will take frequent, brief rests. It does not dig burrows but lives in those made by other species that live in the litter zone. During the breeding season — April to August with a peak in June — two litters of 2 – 8 young are produced. The gestation period lasts for about 22 days, as does the period of lactation. The young at birth are about 0·25 g, and after 14 days are 2·5 g at which they remain until the following year. In mature specimens the head and body is 53 – 63 mm long, the tail 33 – 42 mm, and the weight 2·4 – 5·6 g.

3   **Crocidura russula** (White-toothed or Musk Shrew) is found only on Alderney, Guernsey, and Herm where it occurs abundantly in most habitats.
The breeding season is from March to August with a peak in June. Each litter consists of 3 – 7 young. 'Caravan' behaviour, where the young follow the mother in single file, has been seen in this species. Their lifespan is just over 1 year; most parents die by October and all by January. The head and body is 68 – 80 mm long, the tail 36 – 45 mm, and it weighs 6 – 10 g.

4   **Crocidura suaveolens** (Lesser White-toothed Shrew) is found only in the Scilly Isles, where it is the only shrew, and on Jersey and Sark. Although found in all habitats from a pine plantation to a pebbly beach, most live on heathland. It makes a nest from soft vegetation and feeds upon insects and other invertebrates of the leaf litter and, on stony shores, on sandhoppers. It needs some 7 g of food each day. The head and body measure 57 – 73 mm, the tail is 26 – 38 mm long, and it weighs 3 – 11 g.

5   **Neomys fodiens** (Water Shrew) occurs in England, Scotland, Wales, and on a number of islands. It is an inhabitant of the river bank, where the water is clear, preferably near watercress beds. Its shallow burrow extends along the bank and inland, with a river entrance just above or below the water level. It has regular 'runs' and 'swims' within a home range of some 160 m. It is most active at night. Molluscs, crustacea, and earthworms are eaten as well as small rodents which it kills rapidly with a secretion from its sub-maxillary glands. Snails are soon extracted after the foot has been attacked by the secretion. It partially paralyses earthworms by small bites along their length. Like other shrews it needs much food — some 8 – 12 g each day.
Sexual maturity is reached at 2 years. The breeding season, mid-April to September, has its peak May to June. After about 24 days' gestation, 3 – 8 young, each 1 g, are born. Lactation lasts 37 days, at which time the young weigh about 10 g. A female probably produces two litters a year.
The head and body of mature adults is 76 - 96 mm long, the tail 52 – 72 mm, and the weight 12 – 18 g. During the autumn the long, lustrous winter coat develops and is replaced by a shorter one in spring.

1 COMMON SHREW   2 PIGMY SHREW
3 WHITE-TOOTHED SHREW   4 LESSER WHITE-TOOTHED SHREW
          5 WATER SHREW

VOLES

The animals on this page belong to the order Rodentia, characterized by two large incisor teeth in each jaw. These teeth grow continuously, 2 – 4 mm per week, and as only the front is covered with enamel the soft dentine is worn down to leave a sharp, chisel-like cutting edge. All, except 4, belong to the family Muridae, small, omnivorous animals generally with a naked, scaly tail, and with a high rate of reproduction and rapid development. The coypu, a large animal (4) and a member of the family Capromyidae, is from South America.

1 **Clethrionomys glareolus** (Bank Vole) is widespread on the mainland of England, Scotland and Wales, and also on Skomer, Mull, Raasay and Jersey, each island having its own race.
It inhabits deciduous wood and scrubland, banks, hedgerows, and open terrain where there is cover. It has a well marked territory and a home range of about 40 m in diameter. Juveniles, however, make long distance dispersal movements just before puberty. It lives in a series of tunnels and runs on the surface centred on a burrow, where a nest is constructed of grass. Food consists of green plants, fruits, roots, nuts and fungi, and a small quantity of animal material in May and June. Sexual maturity is reached within 4 or 5 weeks of birth. The breeding period lasts from mid-April to December, 4 or 5 litters being produced in a season. The gestation period is 17 – 21 days, and the young, about 2 g in weight, are weaned in 2·5 weeks. The size of this animal varies depending upon the season. The head and body is 90 – 110 mm in length and the tail 40 – 60 mm, and it weighs 16 – 30 or more g.

2 **Microtus arvalis** (Field Vole). Two subspecies of this vole are found. *M.a. orcadensis* (Orkney Vole) on the mainland of Orkney and on Westray, lives in the litter zone of lowland pastures or in ruined buildings. The head and body is 100 - 124 mm long, the tail 32 - 44 mm, and it weighs 34 – 63 g.
M.a.sarnius (Guernsey Vole), confined to Guernsey, lives mainly in areas of rough grass. The head and body is 83 – 120 mm, the tail 30 – 45 mm, and it weighs up to 45 g.

3 **Microtus agrestis** (Short-tailed Vole) is widespread on the mainland of England, Wales and Scotland, and on some islands of the Inner and Outer Hebrides, the Orkneys, and the Channel Islands. It prefers rough grassland but tolerates most habitats up to 1000 m. A network of tunnels is formed just below the surface — that may cause damage to pastures. It has a small home range and is most active at night although also seen during the day. It feeds chiefly upon grass, eating the bases of the stems. In poor conditions, or in winter, it will eat bulbs, fungi, roots, crops, and even bark. The breeding season is from March to September, and there is a sequence of generations. The gestation period lasts 21 days. The 3 – 6 young in a litter, each about 2 g at birth, are weaned in 14 – 18 days. At 3 weeks, when about 12 g, the females are mature and will mate at 6 weeks. Their measurements vary with the season. The head and body is 90 – 115 mm in length and the tail 31 – 46 mm. They weigh 25 – 30 g up to 40 g. In captivity they have lived for 71 weeks.

4 **Myocastor coypus** (Coypu, Nutria) was imported from South America to breed for its fur, nutria, but it has escaped in East Anglia and Buckinghamshire. It lives in short burrows in river banks, and feeds mainly on aquatic vegetation. It is large and damages crops; efforts are being made to control it. The head and body of an adult is 35 – 58 cm long, the tail 22 – 34 cm, and it can weigh up to 6·4 kg. Two litters of 2 – 9, each about 200 g, are born each year. The young have fur, open eyes and can move around shortly after birth, and take solid food when a day old. They are weaned in 7 – 8 weeks and mature in 3 – 8 months.

5 **Ondatra zibethicus** (Musk Rat), a native of Canada, was imported in 1929 to breed for its fur, musquash. By trapping it was exterminated by 1937.

6 **Arvicola terrestris (amphibius)** (Water Vole, Water Rat) is found in England, Scotland, Wales, and the Isles of Wight and Anglesey. It prefers to live by slow-moving water. A complex tunnel system is built along the bank that includes a nest of reeds or coarse grass and has many exits both above and below water level and others 1 – 10 m away from the water. This territory is marked with musk secreted from the flank glands which are larger and more active in the male. Each scenting site is visited daily and the male will defend his territory. Regular 'swims' exist and paths lead to them. This vole is active throughout 24 hours, spending varying periods above and below ground. It generally feeds near the nest on grass and leaves. Mussels, water snails, fish spawn and worms are eaten and sometimes animals as large as itself.
The breeding season lasts from late March to mid-September; males are fertile in February but females are rarely pregnant before April. There are 5 – 6 young, 7·5 – 10 g at birth, in each litter. Young born early in the season may breed later in the year when about 130 g in weight and 150 mm long, the others breed in the following spring. Few animals, if any, survive a second winter. Males may reach a length of 219 mm for head and body, with a tail of up to 136 mm, and females 220 mm, with a tail of 144 mm. Their weight is very variable but the maximum is about 300 g.

1 BANK VOLE   2 FIELD VOLE

3 SHORT-TAILED VOLE 3A SHORT-TAILED VOLE, young

4 COYPU   5 MUSK RAT   6 WATER VOLE

MICE AND DORMICE

All the animals on this page are rodents (p. 130). Three species (3, 4, and 5) belong to the family Muridae (p. 130), while 1 and 2 belong to the family Gliridae, arboreal, mouse-sized rodents with a bushy tail.

1 **Muscardinus avellanarius** (Dormouse) is indigenous but in the past 50 years there has been a decline in numbers It is found mainly in the south of England and Wales. It prefers to live in the undergrowth of well-established woods and feeds on beech, hazel and sweet chestnut fruits as well as on conifer seeds and the shoots and bark of trees. Arboreal in habit, it will scramble amongst the low bushes and shrubs with great agility.

The dormouse is our only native rodent that hibernates. In the autumn it becomes very fat and retires to its winter nest. There it rolls into a ball, head bent, eyes and mouth tightly closed and ears folded close to the head. As it becomes torpid the respiration and heart rates fall, the muscles become rigid and the temperature drops. Hibernation lasts from September to April but during this period it may rouse and eat some of the stored food.

The breeding period has two peaks, from late June to early July and from late July to early August. The gestation period lasts 21 – 24 days and there are usually 4 young in a litter. They acquire a covering of fur in 13 days and by 18 days the eyes open. They will forage outside the nest at 30 days becoming independent at 40 days. The head and body of the adult is 70 – 86 mm long and the tail 55 – 68 mm. They weigh between 23 and 43 g and are heaviest before hibernation. They may live for 4 years, and have survived 6 years in captivity.

2 **Glis glis** (Edible or Fat Dormouse) was introduced into Tring Park in Hertfordshire in 1902, and has since spread into the surrounding countryside. This species was fattened by the Romans for the table. It is much larger than the native dormouse: the head and body is 175 mm, and the tail about the same; it weighs 150 – 200 g.

3 **Micromys minutus** (Harvest Mouse) is indigenous. It is found mainly in the south and east of England and some border counties of Wales, but is becoming scarce in the midlands and rare further north. It has a more restricted habitat than most rodents. In summer the prehensile tail is useful as much time is spent climbing on the stalks of plants. Winter is spent in burrows just below the surface where they have a nest usually at the end of a tunnel. They feed on fruits, seeds, buds and insects and are active for much of the 24 hours.

A ball-like breeding nest of woven reeds or corn leaves is attached to the stalks (3A). The breeding season is from April to September and the gestation period lasts about 21 days. There may be 5 – 9 young and they develop rapidly; the eyes open in 8 days, at 11 – 12 days they make excursions from the nest, and they are weaned in 15 days when about 5·5 g.

The head and body of the adult is 50 – 63 mm long in the male, 50 – 69 mm in the female, and the tail 46 – 62 mm and 48 – 66 mm respectively. The male weighs 4·2 – 8·2 g and the female 4·4 – 10·2 g. They usually live for 16 – 18 months but can survive for 2 years in captivity.

4 **Apodemus flavicollis** (Yellow-necked Mouse) was introduced and is confined to the southern half of England and many counties of Wales. It occurs in pockets throughout populations of wood mice. It inhabits woods, hedges, and gardens and sometimes buildings. Little is known about this mouse. The beeding season lasts from February to October and an average of 5 embryos has been recorded.

5 **Aopdemus sylvaticus** (Wood Mouse, Long-tailed Field Mouse) is indigenous. These mice are common on the mainland of the British Isles and occur on many of the islands where they are often called subspecies, but their status is not clear.

On the mainland they inhabit woodlands but on the islands almost every habitat. They live in complicated underground tunnel systems amongst decomposing vegetation. A nest of fine grass at the end of a burrow is used as nursery, sleeping space, and food store. Nocturnal animals, they are active throughout the year, and in winter the peak of activity is at dawn and dusk. Although they feed chiefly on seeds, their diet varies with the season. Besides fruit and plant material, much animal matter is eaten between April and July, including the larvae of butterflies and moths. They store food during the winter but their activity is not greatly reduced even in severe weather and indeed they continue to grow throughout the winter.

Most males are fertile by early March when they weigh 15 – 16 g. Pregnant females are found from March till November with a peak in July and August. In some years, breeding continues at a low rate throughout the year. The gestation period is 25 – 26 days; 5 young are usually born and they leave the nest at 15 – 16 days and are weaned at about 21 days. Females may become pregnant within a few weeks of being weaned, when about 12 g. They may live about 6 months, but some survive longer. The population is at its peak in September and October and declines to its lowest in March or April. It is a co-dominant species with the bank vole (p. 130).

1 DORMOUSE 1A part of nest of Dormouse   3 HARVEST MOUSE 3A nest of Harvest Mouse
2 EDIBLE DORMOUSE

4 YELLOW-NECKED MOUSE   5 WOOD MOUSE

RATS AND HOUSE MOUSE

All the animals on this page are rodents (p. 130) and belong to the family Muridae (p. 130). They were all introduced into this country and they cause considerable damage.

1 **Rattus norvegicus** (Brown, Field, Sewer, or Norway Rat) first reached Britain in the 18th century since when it has almost ousted the black rat. It is found mainly in town buildings, yards, and sewers and in farm buildings from which it spreads during summer into fields and woodland, especially where there are rabbit warrens. In a constant environment, such as a hay-rick, breeding may be continuous, but elsewhere there is a peak from March to June, and sometimes a second peak later. They build rather bulky nests from any available material and, after about 24 days' gestation period, 6 – 15 young are born, the number varying with the season. Some intra-uterine mortality occurs in about half of the litters conceived. Some 3 – 5 litters may be produced in a year. The young are weaned and independent at 3 weeks and females are sexually mature when 115 g and about 12 weeks. Males mature when about 200 g.

A wide variety of food is eaten but mainly cereals if available; some 10 – 40 g of wheat may be eaten per day, depending on the rat's size. Animal material is also eaten in summer including insects and crustaceans, and house mice. The head and body of a full-grown male is 203 – 267 mm in length, and of a female 216 – 267 mm; the tail is 165 – 229 mm and 178 – 203 mm respectively. A male can reach a weight of 560 g and a female 540 g. Considerable damage is done by these animals' eating and spoiling grain. They also carry disease organisms including those of food poisoning, *Salmonella*, leptospiral jaundice, and trichinosis.

2 **Rattus rattus** (Black, Ship, or House Rat) was the first rat to arrive in the British Isles, perhaps in the 12th century, and was the only one until the 18th century. In 1964 it was confined to about 28 localities mainly islands and ports. It is a climbing species, often found in top stories of buildings. Black rats are mainly nocturnal, and are omnivorous. The breeding season may continue throughout the year but the peaks are in summer and autumn. They reach sexual maturity at 3 – 4 months. There are 3 – 5 litters per year with about 7 young in each. Gestation lasts 21 days. A full grown male measures 165 – 228 mm along head and body, and a female 143 – 198 mm, the tail being 188 – 252 mm and 140 – 232 mm respectively. Their weight is usually about 200 g. The black rat is a reservoir of bubonic plague particularly in the tropics.

3 **Mus musculus** (House Mouse) has been in Britain since Roman times and has flourished. There is a large urban population associated with houses, factories and warehouses, a rural one around farms, and a small feral one living in fields.

A colony isolated for 70 years has been studied on Skokholm, an island of 100 hectare off Pembrokeshire. These mice are markedly different from others in overall size, number of young in a litter, and skeletal characteristics. They are found in concentrations around buildings and cliffs. Their burrows have a complex runway system, with several branches, chambers, and exits, ending in a circular chamber, 15 cm in diameter. From this home they will move about some 27 m away, occasionally 45 m, but in early spring there are long dispersal movements of 180 m up to 1·5 km. They breed from spring to autumn, the first young being born in April. The average number of embryos increases from 5·0 in March to 10·7 in July, falling to 6·0 in October. The population reflects these changes, reaching an autumn peak of some 4000 – 5000. This number declines and as many as 90 per cent die during the winter. None has been found to survive two winters. On the mainland they breed throughout the year but the number of embryos and of litters produced varies considerably with the environment. An urban mouse has an average of 5·5 litters per year, and 30·9 embryos. In a flour depot a mouse may spend its entire life in a bag of flour. Females in this environment produced an average of 7·9 litters and 44·6 embryos, while those living in a hay-rick had an average of 10·2 litters and 57·3 embryos.

A surprising habitat is a cold store for meat, temperature minus 9·4°C. Here the life cycle takes place in almost total darkness, burrows are made in the meat, and nesting material comes from the sacking encasing the carcasses. The females had an average of 6·7 litters per year and a total of 42·5 embryos.

Mice are omnivorous but little is known of their feeding habits. On Skokholm analysis of mice droppings showed that the majority had eaten mainly plant material, together with arthropods but a few had eaten earthworms.

There is a wide variation in size depending upon habitat and other factors. Those on Skokholm are large and may reach a weight of 50 g. The head and body of males is 72 – 98 mm, and 70 – 102 mm in females, while the tail in both sexes is 70 – 92 mm long. Males weigh 13 – 32 g and females, including pregnant ones, are 10 – 41·5 g.

1 BROWN RAT 1A BROWN RAT. new-born young   2 BLACK RAT
3 HOUSE MOUSE

SQUIRRELS

These members of the order Rodentia (p. 130) belong to the family Sciuridae, characterized by a bushy tail with long hairs. Unlike most rodents, they are active during the day. They cause considerable damage to trees by peeling bark but are useful when they bury seeds and do not recover the hoard.

1 **Sciurus vulgaris** (Red Squirrel) may have been indigenous in coastal parts of southern Scotland and north of the Firths of Forth and Clyde, and it has been in England and Wales from early times but there is no fossil evidence. Carvings exist from the first century and the squirrel has been used as a heraldic device since the 14th century; but the first records date from the 17th century.

The population has shown a number of changes even before the introduction of the grey squirrel (2). In the 18th century there was a decline almost to extinction, perhaps because the forests were first denuded of wood to supply shipyards, furnaces and agriculture, and then burnt. The large-scale replanting after 1750 may have been a factor in the subsequent increase in the population. The red squirrel later became prolific and attempts were made to control it between 1835 and 1933. The numbers dramatically dropped early in this century and many animals showed signs of disease. This decline occurred whether there were grey squirrels in the area or not, but the presence of the grey squirrel appears to prevent the recovery of the red one, though how is not yet clear. The typical habitat of the red squirrel is a coniferous forest, but it will live in mixed woodland areas. The nest or drey, conspicuous in winter, is spherical, about 27 cm in diameter, constructed from twigs and bark and lined with soft material. It can best be seen in winter sited about 10 m up the tree at the junction of a branch and the trunk. There is no visible entrance for the squirrel covers the hole, 5 – 6 cm, as it enters the drey.

The breeding season lasts from January to April but in the south there is a second season in summer. Gestation lasts 42 – 46 days; 1 – 6 young may be born but usually 3. At birth each weighs 10 – 13 g. The eyes and ears are closed until 5 days of age. At 8 days hairs appear on the naked body forming a downy covering until the short dense fur develops by 18 days. Lactation lasts 7 weeks and may continue a further 2 weeks while the young are also sampling solid foods. The lower incisor teeth erupt at 23 days and the upper ones at 41 days. At 8 weeks the young will forage close to the nest. They may be capable of breeding when 6 months old, but this is not certain. It feeds upon acorns, seeds, hazel nuts, hips and haws, and in summer fruit, berries, and greenstuff. Squirrels are hoarders and bury food when it is abundant. Pobably they do not remember the burial site, but rather find it while foraging.

Two moults occur each year. The winter coat, acquired in autumn, is long and soft, the back brownish-grey, the limbs reddish-brown, and the underside creamy-white. The conspicuous ear tufts are lost in late summer or autumn. In spring the summer coat develops, a rich reddish-brown and white or cream underside. The average length of an adult from head to tail is 38 cm, and the weight 312 g; the heaviest one on record was a male of 482 g. The length of life is difficult to assess but red squirrels in captivity have survived for 8 years.

2 **Sciurus carolinensis** (Grey Squirrel) was first introduced from America in 1876, others being imported and released at intervals until 1929. They now inhabit much of England and Wales and a few isolated places in Ireland and Scotland. Its success indicates adaptability in both feeding and dwelling habits. An inhabitant of deciduous woodland, it is also found in parks even in central London. When the leaves have fallen, its drey may be seen against the trunk of an oak tree, which provides its favourite food. It also eats hazel nuts, cones, hard fruits, bulbs, roots, shoots, fungi, soft fruit and grain. Both males and females are sexually mature at 1 year, the first oestrus cycle of the female often occurring at 11 months. There are two breeding seasons: the first begins in the second week of January when mating chases may be seen amongst the leafless branches of the trees; the second begins in the first week of June. Gestation lasts for 44 days. A female's first litter usually consists of 3 young, subsequent ones consist of 5 – 6 young. At birth the young weighs 13 – 17 g and is 10 cm long from nose to tip of tail, the eyes and ears are closed, and the skin is bare. At 2 – 3 weeks hair appears and the lower incisors erupt, followed a week later by the upper ones. At 4 – 5 weeks the eyes open, the molars and premolars begin to erupt, and solid food can be taken as well as milk. Weaning is completed at 8 – 10 weeks. At 7 weeks the young appear outside the nest. If one strays too far, the mother grips the loose skin and fur of the underside with her incisors, while the young one holds her neck and curls his tail around her shoulders. Pre-natal and post-natal mortality appears to be low. This squirrel is heavier and larger than the red squirrel. Its total length is 47 cm; of this the tail forms 20 cm. The average weight is 750 g. The summer coat is short and brownish in colour with a bright rufous streak along the flanks and paws. The winter coat is of long grey hair with a yellow-brown streak down the middle of the back, tapering to the tail, which is fringed with white. The underparts are white throughout the year. Little is known of the length of life but some have survived for 5 years.

1 RED SQUIRRELS
2 GREY SQUIRREL

1A drey of Red Squirrel
2A GREY SQUIRREL.young

THE RABBIT

This animal belongs to the order Lagomorpha, characterized by the presence of four incisor teeth in the upper jaw, two small ones being immediately behind the large ones, and two in the lower jaw. The two large upper and lower incisor teeth grow continuously and have a chisel-like cutting edge. Rabbits are specialized herbivores of medium size that rely upon acute senses of hearing and smell to detect danger, and long hind limbs to escape from it. Rabbits and hares are members of the family Leporidae whose chief characteristic is long ears.

Oryctolagus cuniculus (Rabbit) was introduced in the 12th century by the Normans for food and sport. Rabbits are distributed throughout the mainland and on many islands although it was not until the 19th century that they inhabited the Scottish highlands. They are found on grassland, cultivated fields, and woodlands, and can also thrive on sand-dunes, salt-marshes, mountains, moorlands and cliffs. They are sociable animals and live together in warrens, where an established order of precedence exists. Territories are marked by secretion from the chin gland, as well as with urine and faeces. Warrens are often to be found in the middle of fields where they can extend over 2000 m², and contain as many as 150 rabbits.

As early as November courtship chases may be seen although the breeding season is mainly from January to June. This is determined by the doe being in oestrus, or heat, but the end of the season is due to the decline in potency of the buck. Gestation last for 28 – 30 days. Before the birth of the young the doe prepares a nest of grass formed into a ball, lined with fur plucked from her abdomen. The nest may either be constructed within the warren or a blind burrow, or stop, 30 – 90 cm long may be dug specially. On the island of Skokholm, off south Wales, the doe uses existing, small, blind burrows, 100 – 130 cm long, some of which have been previously dug by shearwaters. The doe prepares the nest and blocks the entrance 2 – 3 days before parturition and returns only when the birth of the young is imminent. At birth the young weigh 30 – 40 g and are blind, deaf, naked and helpless. The doe does not remain with the young but returns, once or twice a day for quite short periods, to suckle them during the first three days after birth. She continues to suckle the young for 3 – 4 weeks by which time they weigh 200 – 300 g and emerge above ground. Sexual maturation is reached in 3 – 4 months; indeed, rabbits born early in the year are capable of breeding in the same season. Although the potential life of a rabbit is about 10 years, in the natural environment the majority probably live for only 1 year. Bucks wait at the mouth of the nesting burrows for the doe to emerge and mating generally occurs within 12 hours of the birth of the young. One litter can be produced each month during the breeding season, the number in each litter increasing to a peak in March and April. The average number of embryos per uterus over

the whole year is 4·4. The reproductive capacity is very high but it is considerably reduced by intra-uterine mortality, in which embryos die before reaching full growth. These embryos are not aborted but are re-sorbed within the uterus, the material being taken back into the body of the mother. This remarkable but quite normal feature in the biology of the rabbit and the hare is not known in any other mammal.

The rabbit eats mainly grass, although under bad conditions it will take bark from saplings. Agricultural and horticultural crops are also eaten. One rabbit can eat about 500 g of fresh green food each day. The feeding ground covers 8000 m² surrounding the warren, the inner 4000 m² being intensively grazed. The rabbit feeds at night, cropping the grass in a semi-circle around its head and stopping frequently to look around for danger from predators. Faecal pellets are deposited at specific sites some distance from the burrow. Rabbits indulge in a habit known as coprophagy or refection, in which the droppings are taken from the vent and passed through the digestive tract a second time. This always occurs during the day while the animal is in the burrow.

The rabbit population was very large though no accurate estimate could be made; in 1948 some 4·5 million rabbit carcasses were sent from railway stations in three counties in west Wales. As one rabbit can eat approximately 180 kg of fresh green food in a year, the total population must consume a very large quantity, of which some at least would be of commercial value. Their control is a problem. In 1954 – 55 the population was greatly reduced by a virus disease, myxomatosis. An animal suffering from this disease is feverish, and the nose and eyes are inflamed and exude a discharge which transmits the disease to other rabbits by contact. There are now signs that the population of rabbits is again increasing. An animal that has recovered from the disease is rarely reinfected because antibodies have been produced that confer immunity. It has been found that an immune doe transfers the antibodies to her foetuses *in utero*.

The head and body of an adult measures up to 40 cm in length and the tail is 4 – 8 cm. The hind limb is large and the hind foot 7·5 – 9·5 cm long. The prominent ears are between 6·5 and 7·0 cm in length. The weight of an adult is 1·150 – 2·000 kg.

1 RABBIT
1A RABBIT young, born almost naked, lying on bed of straw and hair in burrow

HARES

The hare belongs to the order Lagomorpha (p. 138) and to the family Leporidae (p. 138).

1 Lepus timidus (Blue Hare, Arctic Hare) is indigenous but has a more restricted range than the brown hare (2). In Ireland it is found in open habitats and in mountains where there are rock crannies to provide shelter. In Scotland it inhabits open heather moorland and rocky slopes from 300 – 900 m. The blue hare modifies parts of its habitat for different purposes. The most important is a well-marked depression in old heather, rushes or vegetation, called the form, where it lies crouched and concealed. Some forms are used over periods of 10 or more years. Burrows, 1 – 2 m long without branches, are used by the young until they reach adult size. Well-defined communal runs are used regularly and faecal pellets accumulate at stopping places. A dust patch, about 1 m², is used communally; hares spend much time rolling in this and afterwards shake themselves.

In summer they rest in the form for most of the day but at twilight they become active. At first the search for food is interspersed with periods of chasing, rolling, jumping and sand-dusting but subsequently they move to the main feeding area. The chief food is grass and, in winter, heather. Gorse, juniper, birch, pine and soft rush are taken in storms and snow. Refection of food takes place during the morning and afternoon (p. 138). Males are in breeding condition from the end of January until June but females do not reach this condition until February. Pregnancy occurs from the middle of February and the gestation period is about 50 days. Generally all females are pregnant from March until June. On one occasion when copulation was observed 7 hares, including 5 females, scrambled together on an area of 4 m² and four successful matings took place; the group dispersed after 20 minutes. The males do not normally fight even if chasing the same female. Two young are usually born and each female may produce two or three litters per year. At birth the young weigh 71 - 106 g and measure 60 mm or more from crown to rump.

Both sexes are alike in appearance. Three annual moults occur; in spring the colour goes from white to brown, in autumn the brown coat is replaced by one of the same colour, and in winter brown to white. The winter colouration is variable and the majority never attain a completely white coat, the white fur being absent in some animals.

Adults are 45 – 55 cm in length and the tail is 4·3 – 8 cm. The hind foot is 14 cm and the ears about .7·5 cm long. An adult male weighs about 2·6 kg and a female 2·9 kg. They are known to live for 3 – 4 years but some may survive longer.

2 Lepus capensis (europaeus) (Brown or Common Hare) is indigenous on the mainland, where it arrived long after Ireland and the islands had been cut off. It has since been introduced into Ireland where it can be found in a small area in the north-east. It inhabits open ground up to 650 m, moorland, rough pastures, farmland, woodland, and marshes. A solitary animal, it makes its form in tall vegetation and there lies concealed during the day. It has an appetite for crops especially turnips, and will also damage trees by removing the bark. It feeds mainly at dawn and dusk and probably throughout the night.

The 'March madness' of hares is well known and during this time they may be seen chasing and boxing, and often aggregate in small companies. All these activities mark the onset of the breeding season, which is at its peak from May to July. The gestation period lasts for 42 – 44 days and the litter, usually of three young, is deposited in the form. In contrast with the rabbit the young hares, or leverets, are born with their eyes open and completely covered with hair. They can run a day after birth and are weaned in one week, soon afterwards becoming independent. Since it takes 8 months for a young hare to reach sexual maturity, it cannot breed in its first year of life. One female may produce three or four litters each year. Superfoetation has been observed in which conception and development of a second litter occurs before the first litter has been born. As in rabbits there is often a considerable intra-uterine loss of embryos.

The sexes are alike in appearance. There are two moults each year: the first is late in summer or early in autumn when long, dense hair develops that is reddish in colour with grey patches; the second occurs from mid-February to June producing a coat of the same colour but with shorter, less dense hair. The head and body is 52 – 60 cm in length, the tail 8 – 12 cm, the hind foot 15 cm, and the ears 10 cm. An adult weighs between 3·2 and 3·9 kg.

The hare is sometimes hunted by coursing in which it is chased entirely by sight or with harriers, beagles or bassets following the scent.

1 BLUE HARE. summer coat
2 BROWN HARE

1A BLUE HARE. winter coat
2A BROWN HARE. leverets in form

RED DEER

Our only large, wild land animals are members of the order Artiodactyla, even-toed ungulates, whose equally elongated third and fourth digits are protected by hoofs. The deer belong to the family Cervidae whose males are characterized by antlers or tusk-like canine teeth. They are browsers or grazers with cud-chewing habits and a four-chambered stomach, the first chamber being used to store food which is later masticated and digested with the aid of micro-organisms. The speed and highly developed senses of the deer help them to avoid danger.

Cervus elaphus (Red Deer) is indigenous. It is found in many counties and islands of Scotland; in some parts of Ireland; and in several areas of England notably Devon, Somerset and the Lake District, as well as small herds in several other counties. From Norman times it was a protected animal of the royal forests of England until these were destroyed during the Civil War, and was considered the finest beast of the chase. These factors have saved it from extermination. In England it is hunted with horse and hound, while in Scotland it is stalked and shot with a rifle. Its meat, venison, is highly prized. Deer prefer large forests but will if forced move to moorland. They feed mainly on herbage, such as grasses, mosses, or the leaves and shoots of young trees. They will also eat seaweed. In winter most of the day is spent in search of food. In summer the deer feed in the early morning and evening, and remain hidden during the day.

Red deer have a highly developed social structure. For most of the year the sexes are segregated on their own territories. The winter territory of the hinds, between sea level and 500 m, is well defined on three sides but the fourth is variable and leads to the hill tops which provide summer pasture. In their winter territory the hinds calve in June and gather for the rut in October. The stag comes into this territory and will occupy some 10 – 12 hectare, marking it with secretions from the lachrymal (tear) gland and the metatarsal (hind foot) gland, and with urine, and by thrashing trees and bushes. Here he keeps 5 to 50 hinds, depending upon the time of the season. The stag has to work hard to keep the hinds together and guard his territory and as a result may lose weight and condition.

Only the stag bears antlers. They are cast each year early in April and new ones begin to grow almost immediately. At this time he craves minerals for their development and will gnaw shed antlers. By August the antlers are complete and the velvet, a highly vascular skin, is shed towards the end of the month. The antlers increase in size and complexity in successive years, reaching full development in the sixth year. Afterwards there may be some decline in their form. A stag with twelve points is known as a royal. About one stag in a hundred lacks antlers; he is called a hummel, and it is interesting that he is able to gather and keep a harem. Once the velvet is shed, the gonads mature, a mane of hair grows, the neck thickens, the larynx develops, and he can roar when he breaks into rut. The stag is now an impressive creature and during the rutting season is almost black due to frequent visits to the wallows. These are places where the peat can be churned into a creamy consistency by the fore-feet, to a depth in which the deer can roll themselves. Mating is rarely seen but one observer has given the following description. The stag approached a hind who was in season; she ran off and was chased for 20 m; she stopped, approached the stag, rubbed herself along his ribs from fore to hind end, and then attempted to mount him. The stag turned to mount the hind who again ran off and was chased; she stopped, and then the stag mounted and served her. As he was sliding off, the hind kicked her heels and hit him in the belly before running away. The stag roared and followed her again; she came to him and made as if to mount him, he turned and served her again.

Hinds are sexually mature in their third year and produce their first calf usually at 4 years of age, but occasionally at 3 years under favourable conditions. The gestation period lasts about 8 months, a single calf being born in June. Some hinds bear a calf each year, others in alternate years. The average weight of calves, newly born to 1 day old, is 6·7 kg for males and 6·3 kg for females.

Hinds about to calve leave the group and the birth is rarely seen. One report stated that the hind was standing when the calf was dropped. She turned and looked at it for 2 minutes, then moved a few paces away and began to clean herself. Then she returned to the calf, watched for 5 minutes while it lurched to its knees, and then began to lick it. Thirty minutes after birth the calf stood for a few moments. It lay quiet after these efforts and the hind moved about 100 m away and began to graze. The calf usually lies alone for the first few days of its life, the mother visiting it about twice a day to suckle it. Once the calf is able to stand, it follows the hind as she grazes, sucking every few minutes, and the hind fondles it frequently, licks it, and lavishes particular attention on the ears. Maternal care is protracted as lactation extends over 12 months and the young may follow its dam until in its third year. Nevertheless there is a high mortality rate of 50 per cent of calves in the first year. The average lifespan of the species is 15 years. The largest of our deer, the stag stands 105 – 140 cm at the withers, (the ridge between the shoulder blades) and the female about 105 cm. The weight is variable, that of an old stag from a Scottish forest being about 95 kg while one from an English woodland can be almost twice as heavy, 189 kg. The antlers are about 100 cm long with a span of 75 cm and they weigh 7 kg.

1 RED DEER, stag   1A RED DEER, hind   1B RED DEER, calf

ROE DEER

The roe deer belongs to the order Artiodactyla (p. 142), and to the family Cervidae (p. 142).

Capreolus capreolus (Roe Deer) is indigenous and is found throughout Scotland, some parts of northern England, and several counties in the south including Dorset, Devon, and Sussex. Deer parks, which have existed since the Norman conquest, were formed by noblemen to indulge their fondness for hunting and to provide fresh meat. Until early Tudor times large areas of England were royal forests, where only the king and his guests could hunt. The roe was one of four beasts protected by forest laws relating to venison.

Roe inhabit the edges of forests, open wood, or scrubland, that provides thick cover for them to lie up in during the day. They are active mainly at night, browsing on the leaves of willow, birch, oak and bramble or eating grass, berries and fungi.

The antlers are a conspicuous secondary sexual characteristic of the buck, grown and shed annually, they are lost in November and December, and a few days afterwards a growth of velvet (see p. 142) laps the scars and new antlers form rapidly. By early May they are complete. The size and complexity of the points, or tines, may increase until the animal is four years of age, when they reach their peak. The buck uses them in threat and fighting when they act both as weapon and shield, and in territorial marking, as well as for scratching himself.

In spring the buck cleans his antlers of velvet and begins to take up his stand, or territory, which he will defend during the rut. The boundaries of the territory, about 5 hectare, are marked by 'scrapes' on the bases of trees and shrubs, formed by a backward flailing movement of the foreleg. The same scrapes will be used over long periods even though the territory is occupied by a different buck. Fraying stocks, made by the movement of antlers up and down the stem of a tree to remove the velvet, are marked by secretion from the glands which lie in front of the bony stalks, or pedicles, of the antler. Development and regression of this gland follows an annual cycle similar to that of the testis, the function being associated with reproductive behaviour. The buck wipes the secretion on to the fraying stocks throughout the summer, presumably to define the boundaries. Another feature is the rutting or racing-ring, formed around some natural object such as a bush or tree which is also used for many years.

During the rutting season which lasts from June to the end of August the bucks are aggressive and restless. Both buck and doe make characteristic calls. The doe has a high-pitched rhythmical cry which attracts bucks from considerable distances, while the buck makes a deep, loud rasping cry, generally when in pursuit of the female. As the rutting season approaches, the doe indulges in leaps, jumps and short dashes. From the onset of this behaviour the buck attends the doe closely and courtship follows. The buck chases the doe for quite long distances and they have been seen to walk round a rutting ring for half an hour, the doe being mounted or served by the buck nine times.

There is a delay of some months between fertilization, which takes place during the rut, and the implantation of the embryo on the wall of the uterus which takes place in December. The gestation period continues for 5 – 6 months and the young, generally twins, are born in May or the beginning of June. At this time the doe retires by herself after driving off the young of the previous season. Young does mature early and may become pregnant at 15 months and give birth to the next generation in two years. A buck of two years, however, cannot compete with older bucks and will not usually mate until 3 years.

The coat of a kid is pale brown flecked with white, providing excellent camouflage reinforced with its ability to lie without movement for long periods while the mother is absent on a foraging expedition. The coat soon fades to a sandy red which remains until the first adult coat is assumed in October. Most of the does have their thick winter pelage by mid-October. It varies in colour from brown to almost black, shading to yellow or white on the underside. There is a large patch of white perianal hairs — oval in the male, heart-shaped in the female — that forms a distinguishing feature between the sexes after the bucks have shed their antlers. The winter coat begins to fall out in the middle of March, the last remnants remaining sometimes until the end of May. The summer pelage is short and sandy to deep red-brown in colour and the perianal region is buff. The colour of the fur deepens as they reach their prime. The roe is the smallest of the indigenous European deer. A mature buck weighs approximately 23 – 26 kg and a doe 11 – 21 kg. Height at the shoulder is approximately 65 – 73 cm, while the length from nose to tail is 120 cm. The antlers are about 22 cm in length.

ROE DEER

1 buck   1A doe   1B kid

FALLOW DEER

The animals on this page belong to the order Artiodactyla (p. 142) and to the family Cervidae (p. 142).

1 Dama dama (Fallow Deer), indigenous in southern Europe and Asia Minor, was introduced into Britain by Phoenicians or Romans and recorded in the Domesday Book as well established. In Norman times they were numerous and highly regarded, kept in parks for the hunt and for venison. They were present in Hyde Park in the year of Queen Victoria's coronation. Through the ages many have escaped from parks and today are widespread in England and Ireland, found in fourteen counties and three islands of Scotland, but scarce in Wales.

They inhabit lowland woods, preferably deciduous, with thickets for daytime cover and access to pastureland. They eat a wide variety of food, mainly grasses, herbs, shrubs, and fruit. They feed during the evening and night, especially moonlit nights, and also at first light as they move to their daytime resting place. In winter feeding continues during the day.

Bucks are found in herds from the beginning of the rutting season in autumn until they begin to cast their antlers when they become solitary and secretive until the new ones have grown and the velvet is cleaned. Antlers, present only on the male, are different from those of other British deer. They have a broad palmate area, which is fully developed at 6 or 7 years. Each year new antlers begin to grow immediately after the old ones have been lost in April. Their development takes about 4 months and they are cleaned of velvet by the beginning of September in preparation for the rutting season. A young buck fawn develops bumps which become pedicles by the spring following his birth. Two spikes about the length of the ear develop from these pedicles, and by late August are cleaned of velvet. Each year for 5 years the antler becomes more complex, evidence of the deer's age. A yearling is a *pricket*, one of 2 years a *sorel*, one of 4 years a *bare buck*, one of 5 years a *buck*, and one of more than 5 years is a *great buck*.

The rutting season lasts from late September or October until the beginning of November. Preparation for the rut is indicated in the buck by a swelling of the neck, humping of the back, and cessation of feeding. He establishes a territory around the perimeter of the rutting stand by thrashing young trees and bushes, and annointing them with a secretion from glands just below the eyes. Many rutting stands have been used over long periods of time, the buck returning to the same stand year after year. At this time he produces a sound made only during the rut, a continuous throaty grunting. He will parade up and down his rutting stand, groaning for many hours to attract does in season which are served by the buck, often after a long chase. Fights occur when other bucks enter the territory.

Does do not usually breed until they are about 16 months old. The young are born in late May or early June, usually only one fawn for twins are rare. Just before the birth the doe becomes secretive and remains in dense cover. After the birth she may be seen alone, often at some distance from the fawn. Once the fawn is able to follow its mother, two or three does with their young can be seen together until the following March. Then the doe is again bearing young and weans the one that has been with her since birth. Nursery parties of fawns herding together may be seen at this time.

The summer coat of the adults is reddish brown with white or yellowish white spots. Along its flanks is a line of white hairs, while there is a line of black hairs down the spine and tail. The rump is white outlined with black, and the underparts are white. In winter the coat becomes a uniform greyish brown, or black, shading to a paler brown on the lower flanks and white on the belly. Both summer and winter coats provide excellent camouflage. The fawns have on their coat large white spots. Besides this common colouration, there are at least three other colour varieties, namely menil, black, and white.

The senses are highly developed although the least keen is sight; they are able to see movement, but a stationary man is rarely observed. Their hearing is acute. Most efficient of all is their sense of smell and they seem to be able to pinpoint accurately the position from which the scent comes. In general, this species relies rather on its ability to conceal itself from danger than on speed to escape from it.

The buck stands 80 – 90 cm at the shoulder and the doe 80 – 85 cm. The weight varies with the locality, a great buck averaging 77 – 90 kg, and a buck 63 – 68 kg. The doe is much smaller and weighs 31 – 36 kg. At birth the fawn weighs about 4 kg.

2 Cervus nippon (Sika or Japanese Deer) were presented to the Zoological Society of London in 1860 and to other parks. Feral herds now occur in England, Ireland, and Scotland in deciduous or mixed woodland with brambles and thickets.

Closely related to the red deer (p. 142), its life cycle and behaviour are similar. Stags and hinds remain apart except in the rutting season in September and October. Their antlers, shed in April or May, develop during spring and summer, generally with four points and a length of about 50 cm. The velvet is shed early in September as the rut approaches when the stag wallows and produces a penetrating whistle. A single calf is usually born in June. The stag is about 90 cm at the shoulder, the hind 70 cm, and they weigh 45 – 64 kg and about 36 kg respectively.

1 FALLOW DEER, buck   1A FALLOW DEER, doe   1B FALLOW DEER, fawn
2 SIKA DEER, stag   2A SIKA DEER, hind   2B SIKA DEER, calf

REINDEER, MUNTJAC AND WATER DEER

All the animals on this page belong to the order Artiodactyla (p. 142) and to the family Cervidae (p. 142).

1 Muntiacus muntjak (Indian Muntjac) was introduced in 1900 by the Duke of Bedford. This species is found only in zoos and parks, but some animals exist which are a cross between the Indian and Chinese muntjac and are intermediate in size between the two species.

2 Muntiacus reevesi (Chinese Muntjac, Barking or Rib-faced Deer) was introduced from China, in 1900. The muntjacs are small with simple antlers, supported by long skin-covered pedicles which continue down the forehead as convergent ridges marked by dark lines of hair. This feature gives rise to the name, rib-faced deer. They also have tusk-like canine teeth which in the bucks are curved and about 2 cm long. The other feature of this deer is the loud dog-like bark, uttered at regular intervals when the animal is alarmed or disturbed. The buck stands 43 – 46 cm at the shoulder and the doe 38 – 40 cm; and they weigh about 20 kg and 13 kg respectively.

3 Hydropotes inermis (Chinese Water Deer) was introduced early this century by the Duke of Bedford and it has since become established and multiplied rapidly. Although there have been escapes it has only spread a few miles around Woburn Park. In China it inhabits broad-leaved woodland zones by the shores of rivers and lakes. In England it has lived successfully in dry woodland.

A solitary grazing species, little is known of its social or breeding behaviour. The rut occurs in December when the male will emit a loud whistle. The young are born in June or July, generally twins or triplets although up to six have been recorded. Antlers are absent in both sexes but the upper canine teeth form curved tusks, about 7 cm long in the male and a little shorter in the female. The buck is about 50 cm at the shoulder and the doe 48 cm; they weigh some 12 kg and 10 kg respectively.

4 Rangifer tarandus (Reindeer) last inhabited Caithness in the north of Scotland in the 12th century. Man probably restricted its range by destroying the forest floor by fire and finally brought about its extermination by hunting.

In 1952 Swedish mountain reindeer were re-introduced into the Cairngorm mountains of Inverness-shire. Later, to prevent inbreeding, these animals were joined by Swedish woodland and South Norwegian reindeer. The herd was given pastures in the Cairngorms overlooking the river Spey, and more recently have grazed in Glenmore Forest Park and to the summits of Cairn Gorm and Cairn Lochain. Reindeer inhabit regions where ground, rock and tree lichens are found. Lichens form their chief food including *Cladonia rangifer* or 'reindeer moss' as well as other species. Reindeer live in herds consisting largely of females and young, generally led by an old male. Mature males are solitary in the summer, joining the herd only for the rutting season. Although these animals are domesticated, they resemble their wild ancestors more closely than do most domesticated animals. The Scottish reindeer are controlled by the voice of their herder whereas in Lapland dogs help in herding. The unique feature of the reindeer is the presence of antlers in both sexes. These are large in the male, or bull, sometimes reaching 125 cm in length, with great diversity in shape. A brow tine consisting of a large palmate shovel is generally present on one antler only. The top points are frequently palmate and a back tine is present about halfway along the beam. Bulls shed their antlers in December but new ones do not grow until the spring. The females have smaller and somewhat simpler antlers and do not shed them until after the birth of their young.

The rut occurs in September and October. At this time the strongest male will hold a herd against intruders and may be heard roaring. A single calf is born in May or June, after a gestation period of about 8 months. It is fully weaned by the time of the next rut and sexually mature by the autumn of the following year.

The coat of the adults varies in summer from a dark greyish brown to almost white, with white underparts; in winter it is grey-brown. The newborn calf is brown, tending to be darker along the back, and is without spots.

The bull is 112 – 120 cm in height at the shoulder, and weighs 120 – 150 kg, while the cow is slightly smaller being 92 – 102 cm in height.

1 INDIAN MUNTJAC, buck
1A INDIAN MUNTJAC, doe

2 CHINESE MUNTJAC, buck
2A CHINESE MUNTJAC, doe

3 CHINESE WATER DEER, buck 3A CHINESE WATER DEER, doe
4 REINDEER, bull 4A REINDEER, cow 4B REINDEER, calf

SHEEP, CATTLE AND GOATS

These animals belong to the Bovidae, the largest family of the order Artiodactyla (p. 142). This family is characterized by horns which are always present on males and also on females in many species. These are permanent, continuously growing, unbranched, bony outgrowths encased in a sheath of horn. Many members of this family have been domesticated.

1 **Capra hircus** (Goat) is found on the Snowdon and Cader Idris groups of mountains in Wales, and in mountainous areas in the north of England, Scotland and Ireland, as well as on the islands of Lundy and Skokholm. It has probably been present in some places since escaping from domestic herds of Neolithic settlers, but has been supplemented by later escapes. The goat, one of the oldest domesticated farm animals, has been associated with man for some 10,000 years. Goats were kept by cottagers for milk until the end of the 18th century, but there was a decline after the industrial revolution, although some are still kept today particularly in isolated areas.

The feral goats live in small, loosely associated groups of 3 – 10 females, with one or more males during the rutting season. At this period they are less wary of intruders and there is considerable activity. The males chase the females and this is interrupted by sporadic fighting. Fights between males, often spectacular, occur most frequently early in the season. The males draw back and then charge with lowered heads, time after time, often for an hour or more.

Mating occurs during October and November and the gestation period lasts about 150 days. Usually the kid is born in March, although some are born in February. Occasionally twins are born but their mortality is even higher than that of singletons. A fully grown male weighs 45 – 56 kg including the head with the massive horns, and a female weighs 22 – 31 kg. The scimitar-shaped horns bear distinctive rings which reveal annual changes in their growth and can be used in determining age. Those of a 3-year-old male are 39 cm along the curve and those of a 9 – 10 year one are 76 cm.

Goats eat a wide range of food and being able climbers can feed in places which are inaccessible to other animals. They will move down to be below 500 m to shelter during wet, cold or windy weather.

2 **Ovis aries** (Soay or Wild Sheep) was probably introduced into the island of Soay, in the St. Kilda group, about 1000 years ago by the Vikings. It is the most primitive domestic breed in Europe and resembles sheep of Neolithic times. The Soay sheep now live also on a neighbouring island, Hirta, where they were introduced in the 1930s. They shelter in 'cleits', dry stone cells built by the St. Kildans to store peat, hay, and sea-bird carcasses. Much of their time is spent in groups, aggregating in the evening into small flocks to climb to higher ground on the periphery of their home range. There is a reverse, and often rapid, movement at dawn to lower ground from which they scatter. This daily cycle of movements varies with weather and with season. Each flock consists of both rams and ewes, one flock being found to include some 12 rams. They eat the natural vegetation of grassland and heath.

The mating season probably lasts from early October to late November or early December. Lambing starts early in March, a peak in the birth rate occurring in the last week of March and the first week of April, declining towards the end of April. The majority of lambs are singletons, but twins are not uncommon. At birth a lamb weighs 924 – 2520 g and its coat is very dark brown in colour. It is active immediately after birth and will keep up with its mother when only a few hours old. If an intruder appears, the mother will move away with the lamb on the far side. In most years there is a high mortality of males in their second winter, but no corresponding loss occurs in females of the same age.

These sheep are small; the male is 52 cm long, the female 49 cm, and they weigh about 25 kg and 19 kg respectively. The coat is of short, fine, soft wool which tends to rub off in late winter and early spring. They can live for 5 years or more.

3 **Bos taurus** (Chillingham or White Cattle) were first enclosed 700 years ago in Chillingham Park, Northumberland. This herd alone is completely pure and in 1972 it numbered 48. Three yearlings — 2 heifers and a bull — were then withdrawn from the herd as a basis for a reserve herd to be established on secluded Crown land in the north of Scotland. Although the origin of British White cattle is not known, they were plentiful in forests before the Iron Age. They may be descendants of Celtic short-horn cattle which were introduced in Neolithic times.

They have not become domesticated during their enclosure and they dislike the scent of man. A leader, or king bull, is selected by combat and he reigns until deposed by a younger and stronger rival. The vanquished bull is banished from the herd. Until 1826 men hunted and killed him but now there is no human interference. The females seek seclusion when about to drop their young and remain away from the herd for about a week. These cattle live on natural vegetation but are given some hay during winter. It is said that they dislike supplementary cattle food although some individuals will taste silage (fermented forage).

The average live weight of a mature bull is about 600 kg and of a cow 450 kg.

1 GOAT. billy   1A GOAT, nanny   1B GOAT, kid
2 SOAY SHEEP. ram   2A SOAY SHEEP, ewe   2B SOAY SHEEP, lamb
3 CHILLINGHAM CATTLE . bull   3A CHILLINGHAM CATTLE, cow   3B CHILLINGHAM CATTLE, calf

PONIES

The indigenous wild horse was the only European representative of the order Perissodactyla, the odd-toed ungulates, but it has been exterminated throughout most of its range. The horse, *Equus caballus*, belongs to the family Equidae, specialized for swift movement, the weight being carried by the central digit. The dentition is highly developed for eating grasses, the incisors for cropping and the molars for grinding tough food. It is not known whether there was an indigenous wild horse in the British Isles or whether domestic breeds were introduced from the continent. Modern 'wild' ponies, owned by farmers but pastured on common grazing lands, breed without attention and are only occasionally rounded up.

1 **Shetland Pony** or **Sheltie.** This is the smallest of our ponies being about 101 cm high; the smallest one on record was only 66 cm high. The colour may be black, brown, bay, grey, dun, or chestnut, and piebald and skewbald ones also occur. In winter the coat is of long hair with an undergrowth of wool. Until they are 2 years old ponies have a coat of wool rather than hair. Extremely hardy, they can survive on sea-weed picked off the shore if conditions are very bad. The Shetland pony is a useful draught animal, able to pull a very considerable load. Many survive for 30 years and some for 40.

2 **Dartmoor Pony.** These ponies have remained on Dartmoor both before and during its development from a deer forest to a grazing area. They run free on the moors and are handled only for branding. They stand about 125 cm high and may be bay, black or brown.

3 **New Forest Pony.** An early description of this Hampshire pony was of a stocky animal 111 – 121 cm high, hardy and able to survive the winter unattended. Until the mid-19th century it was used as a pack or colliery pony. Arab stallions were introduced to improve the stock in the 1850s, the 1870s, the 1880s, and early in this century. This increased the pony's height and slenderness but made it less hardy. The present-day pony is large being 121 – 142 cm high. The colour varies but bays and brown predominate. The ponies graze mainly in the valley bottoms and on acid grassland, preferring purple moor grass to bristle bent. During the winter many of them feed off gorse, holly, bilberry and ivy as well as saplings of many hardwoods.

Connemara Pony. Found in Connacht to the west of Loughs Corrib and Mask, in Ireland, these ponies subsist on the scanty rough herbage of mountain, bog and shore, surviving under bad conditions. They are said to be very fertile and free from hereditary diseases. Mares have been known to survive 20 or more years. Late in the 19th century Arab blood was introduced. The colour of this pony is very variable; the majority are grey but there are black, bay, brown and dun ones with occasional chestnut and roan ones, the latter having an admixture of white that lightens the main body colour. The body is compact and deep, the legs short. Their height is usually between 132 and 142 cm.

Exmoor Pony is dark brown, the nostrils glossy black and the muzzle the colour of fine meal. The same mealy colour surrounds each eye, and is inside the fore-arms and thighs, as well as under the belly.
Quite a small pony, it is about 127 cm high when 4 years old. The larger ponies were used in breeding the west of England pack horses. Today most of the ponies belong to farmers. The mares foal over a long period, from March to November, but the peak occurs in April and May.

Welsh Mountain Pony is usually grey, bay or brown in colour but black, roan, chestnut, cream and dun ones are also found. It may be up to 121 cm high, The Welsh Cob is of relatively recent origin and is similar in colour to the mountain pony but stands up to 151 cm high.

Fell Pony is found in Cumberland and Westmorland to the west of the Pennines. It is very similar to the Dales pony and is usually black or black-brown in colour and stands some 132 cm high.

Dales Pony is found in the bleak northern uplands of the upper reaches of the Rivers Tees and Wear in County Durham, and of Westmorland and Northumberland east of the Pennines. It is used on hillside farms and will carry very heavy loads. It eats the rough grass of poor hill land and takes only the minimum of forage. Standing some 145 cm high, it resembles a miniature cart-horse, the large size being due to interbreeding probably with Clydesdale horses. Many are jet black but bay, brown and grey ones may be found.

Highland Pony has descended from wild ponies of the highlands of Scotland but has been crossed with Norse blood and later Arab blood. It is 127 – 147 cm high at the shoulder. It may be black, brown or fox with a silver mane and tail, or dun or grey with no white markings. An eel stripe along the back is a typical feature but is not always present.

1 SHETLAND PONY
2 DARTMOOR PONY
3 NEW FOREST PONY, and foal

HORSESHOE BATS

Bats are members of the order Chiroptera ('hand-winged'), the only group of mammals capable of sustained flight. The characteristic feature is the wing, a double fold of delicate skin attached to the elongated bones of the arm and hand and along the length of the body extending back to join the ankle and tail. Usually active at dusk or during the night, they avoid obstacles by echolocation, the emission of sounds of short wave length that are reflected back to the animal by objects in the flight path. Some of these sounds have a component audible to man. Most bats undergo hibernation. The two bats on this page belong to the family Rhinolophidae, easily recognized by the prominent, horseshoe-like growth on the nose.

1 **Rhinolophus hipposideros** (Lesser Horseshoe Bat) is indigenous and has a slightly wider distribution than the greater horseshoe bat. It is found in the west of Ireland, in Wales, and in south-west England.

They are generally found in caves hanging singly in sheltered crevices or beneath projecting rocks only a short distance above the ground. They hang upside down by the feet so that the body is free from contact with the surrounding rock. In the summer they may live in colonies; one colony found in the roof of a house consisted of about thirty adults, almost half of which were males.

They feed chiefly on moths, but also eat small beetles and spiders. In flight this species remains close to the ground and a high pitched sqeak may be heard. Flights, which occur in summer, are over distances of at least 1·5 – 3 km; one recorded was between places 22 km apart. Hibernation begins in October but until December flights within their hibernation cave occur, when they will also feed. From December to March hibernation is quite deep. The pattern of reproduction is similar to that in the greater horseshoe bat (2) and a single young of about 2 g is born in June or July.

This is one of the smaller of the British bats and except for size it closely resembles the greater horseshoe bat. The head and body of the adult is 35 – 39 mm long, the forearm 37 – 40 mm, and the wing-span 228 – 250 mm. The weight of the adult, which is variable because of hibernation, ranges between 4·3 g and 6·8 g. An age of 12·5 years has been recorded.

2 **Rhinolophus ferrumequinum** (Greater Horseshoe Bat) is found mainly in south-west England and south Wales. Groups have been studied in Devon, Somerset, Dorset and Gloucestershire.

During the summer they are difficult to find but solitary individuals are occasionally discovered. In winter colonies of 6 to several hundred gather together in a cave or mine where they may be loosely spaced with small gaps or else so tightly packed that it is impossible to distinguish individuals. It can be recognized at the entrance to a cave or mines, as it is the only one of the three large British bats to hibernate in such a habitat. In flight it remains close to the ground, generally within 1 m, and sometimes lands to pick up insects. Flights of at least 1 – 3 km from the cave are often made both in summer and winter, the longest recorded flight being one of at least 64 km. It has a loud, high-pitched, penetrating squeak.

Food consists mainly of beetles, such as the greater dor beetle, the dung beetle, and the cockchafer, but also of other insects, including crane flies and moths. Smaller insects are eaten during flight; larger ones are taken to a feeding post, a branch of a tree, a porch, outhouse, or tunnel, to be dismembered. Feeding generally takes place at dusk, with flights at intervals throughout the night.

Hibernation is not deep in this species. It lasts from late September until mid-May, and during this period flights may be made either in pursuit of food on warm days when insects are about, or to reach a cave where better conditions of temperature exist. The temperature of the bat becomes quite low in hibernation, and indeed in any period of rest, and consequently it is important to find a suitable site. It tolerates a site where the temperature is 5 – 12°C but prefers 7 – 9°C. When the horseshoe bat is sleeping or resting, the wings completely enclose it. Breeding occurs in the female at the end of the third year, and afterwards a pair of false teats, used by the infant for anchorage to the mother during flight, are present in the groin. Copulation takes place in October and the live sperm are stored during the winter until April when ovulation and consequent fertilization takes place. In July females gather together to form a 'nursing colony', which can be large; one contained 150 adults among which were several males. A single young is born in July; it weighs about 6·8 g, is blind, and has a covering of soft, silky hair. Growth is rapid and by 40 days the bones of the fore-arms have reached their maximum size. When the mothers are feeding, the young remain in the colony in loudly squeaking clusters. Although breeding does not occur until the third year and only one offspring is produced, the bats survive for a number of years. The greatest recorded age is 19·25 years. The head and body of the adult is 65 – 68 mm long, the fore-arm 54 – 58 mm, and the wing-span 330 – 385 mm. During hibernation 23 to 30 per cent of the peak weight is lost; the minimum weight is 13·4 g the maximum 29 g, females being larger than males.

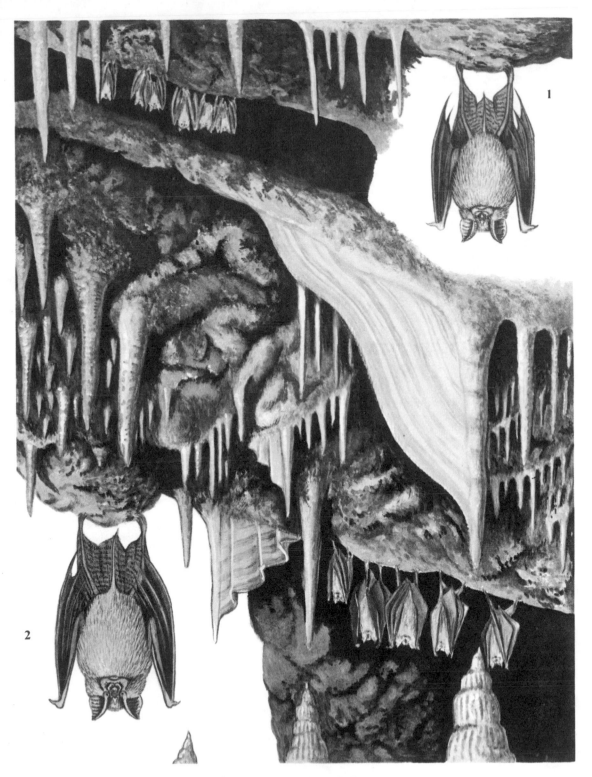

1 LESSER HORSESHOE BAT
2 GREATER HORSESHOE BAT

BATS

Bats belong to the order Chiroptera (p. 154). All those bats found in the British Isles, except the horseshoe bats (p. 154) are members of the family Vespertilionidae, characterized by a nose without modifications, a growth from the inner margin of the ear (the tragus), and a well-developed interfemoral membrane supported throughout its length by a fairly long tail. In the majority of species and in all the British ones the female has one pair of teats.

1 **Myotis nattereri** (Natterer's Bat) is probably widespread in most of England, Wales and Ireland, and a few records exist from Scotland in the 19th century.

It is gregarious and some colonies reach huge proportions. It lives in trees and buildings, and likes heavily timbered land near water. From the end of September to March it hibernates although on mild nights it will often rouse from hibernation and become active for short periods. In summer it emerges early in the evening and flies at intervals throughout the night. Its flight is slow and steady with the tail held straight out behind, which makes it easy to recognize. It is usually to be seen flying around trees at a moderate height while feeding chiefly on moths. Some prey is picked off the foliage. While hunting it produces a shrill, continuous squeak which can be heard by man.

Copulation has been observed in December. A single young, blind and naked, is born at the end of June and by the middle of August is capable of flight. The head and body of the adult is 43 – 50 mm long, the wingspan 265 – 285 mm, and the weight 5 – 10·3 g.

2 **Myotis mystacinus** (Whiskered Bat) is found in isolated areas throughout England, in the border areas of Wales, and in Scotland.

The summer is spent in trees or buildings. It hibernates from the end of October to the middle of March in caves, rock fissures, old quarries, or buildings. A small number were found hibernating in an underground passage together with *Myotis myotis* (3), *M. daubentoni* (p. 158), *Myotis nattereri* (1) and *Plecotus auritus* (p. 160). During summer it emerges early in the evening, often before sunset, and is active for much of the night. It may sometimes be seen in the day especially in spring. Flies and gnats are taken on the wing but it prefers spiders, which it picks from foliage or fences. The

flight of this bat is slow, often hovering, and low, 1 – 4 m. During flight the underparts appear quite white. It can produce a low buzzing squeak. Copulation occurs in December. A single young is born, blind and naked, in June or July and by the beginning of August it is able to fly.

The head and body of the adult are 46 – 50 mm long, the wingspan 225 – 245 mm, and the body weight is 4·5 – 6 g. A male of 7 years has been recorded.

3 **Myotis myotis** (Mouse-eared Bat) is a common species in Belgium, France, Germany, Holland and Switzerland. It was first recorded in this country in 1888, in Girton near Cambridge. There are a number of recent records from 1956 to the present time from Dorset and Sussex.

Nine were found in an underground passage in Sussex during October. They were hanging by feet and thumbs some 3·5 m from the ground. Of these 7 were solitary; the remaining two were hanging together, the male clinging to the back of the female. In November they were still present and by the end of January they had been joined by other myotids. Their surroundings had a high relative humidity and the bats were torpid and covered with drops of condensed moisture.

During summer this bat emerges late from its roost and flies slowly along a straight course at a medium height. A flight may last for 4 – 5 hours. This species undertakes considerable migrations, and one of at least 260 km has been recorded on the continent.

The head and body of the male is some 78 mm and of the female 80 mm, the wingspan about 368 mm and 394 mm, and the tail 49 mm and 46 mm respectively. It weighs 18 – 45 g and is the largest of the European species. On the continent it has been kept in captivity for up to 2 years.

1 NATTERER'S BAT 1A NATTERER'S BAT. wing membrane, between hind feet, edged with stiff short hairs
2 WHISKERED BAT   3 MOUSE-EARED BAT

BATS

All the bats on this page belong to the order Chiroptera (p. 154) and the family Vespertilionidae (p. 156).

1 Pipistrellus nathusii (Nathusius's Pipistrelle) was first found in this country in Dorset in October 1969 in the roof of a research station. It is regarded as uncommon or rare in western Europe, although well documented from Denmark, Germany, Poland, Switzerland, Italy and eastern Europe.

The overall colouration is grey but the ventral fur is light and the wing has an ill-defined light border. The head and body of the male found measured 50 mm, the wingspan was 242 mm, and it weighed 6·40 g.

2 Myotis bechsteini (Bechstein's Bat) has been recorded from Berkshire, Dorset, Gloucestershire, Hampshire, Shropshire, Somerset, Wiltshire and the Isle of Wight. In Dorset it has been found hibernating in caves, hanging from the roof of a tunnel, and also hanging free in crevices little larger than itself. One female remained in the same position for 56 days. A large preponderance of males to females was found in the winter colonies, the ratio being 9 to 4.

They live in small colonies in hollow trees during the summer. Shortly after sunset they emerge and fly rather ponderously, at a height of 1 – 4 m, with their large ears held straight out in front. When in flight they make no audible sound but are capable of producing a low buzz and a high-pitched squeal. They probably feed mainly upon moths.

The head and body of the male is 45 – 50 mm and of the female is about 43 mm, the wingspan is about 270 mm and 263 mm respectively; their weight is 9 – 11 g. The oldest one recorded was a male of at least 15 years.

3 Myotis daubentoni (Daubenton's Bat, Water Bat) is found in England, north Wales, the southern half of Scotland, and the eastern half of Ireland.

Oak woods in close proximity to water are a favourite habitat but they will also live in caves, rock fissures and buildings in summer. Colonies of several hundred bats of both sexes are often found. In winter they will hibernate in buildings and caves.

During the summer they emerge from their roost in rapid succession, late in the evening, after a prolonged period of squeaking. They hunt over water, often in large numbers, flying in wide circles low over its surface. They can be observed throughout the night until about 15 minutes before sunrise, although each bat returns frequently to the roost. In flight they make no audible sound. Mayfly are their chief prey but other large insects including moths and dragonflies are also eaten. In June or July a single young is born blind and naked. By late August it is capable of flight. The head and body of the adult is 46 – 51 mm, the wingspan 220 – 245 mm and the weight is 7 – 11 g. The oldest recorded specimen, found in Suffolk, was 18·5 years of age.

1 NATHUSIUS'S PIPISTRELLE
2 BECHSTEIN'S BAT
3 DAUBENTON'S BAT

BATS

The bats on this page are all members of the order Chiroptera (p. 154) and the family Vesper-tilionidae (p. 156).

1 **Vespertilio murinus** (Parti-coloured Bat) occurs only rarely; two records exist from before 1835, another in 1927, and one was captured on a drilling rig in 1966, in the North Sea 270 km due east of Berwick.
It usually lives in hollow trees or in buildings. The head and body is 56 – 63 mm long and the wingspan 260 – 286 mm.

2 **Barbastella barbastellus** (Barbastelle) is an elusive species found mainly in the southern half of England with isolated records from Yorkshire, Carlisle, and Wales.
It is solitary in habit except in August and September when small colonies may be seen in the day. It hangs in the crevices of tree trunks, in thatch, or behind shutters. Emergence from the roost begins at sunset and lasts for about 30 minutes. Activity continues until midnight in August and September, and there is a further period of activity before dawn. It flies at a height of 1 – 4 m and as it nears a resting place flight is slow and uncertain. Although it is usually silent when flying, a deep buzz and a series of harsh squeaks can be produced. Whilst flying it feeds upon insects, chiefly moths, flies and gnats. From the end of September to mid-April or even May it hibernates in roofs of buildings.
A single young is born in July that is blind and naked but by the middle of August is capable of flight.
The head and body of the adult is 49 – 52 mm long, the wingspan 254 – 268 mm, and the weight 6 – 8 g.

3 **Plecotus auritus** (Common Long-eared Bat) is wide-spread throughout the mainland of England, Wales, Scotland and Ireland.
The sexes roost together in colonies throughout the year although some have been found alone. They live in trees and buildings during the summer but the nursing of young may occur in a different area from that usually occupied by the colony. In winter they hibernate in buildings and caves from November to March.
Males may reach sexual maturity in their first year while females may produce their first young in their second year of life. The main period of mating takes place in May but it may also occur in October and November. A single young is usually born in July; some born as early as May were found not to survive. The newly born young weighs 0·9 g and has a wingspan of 75 mm.

The adults feed chiefly upon moths and vanessid butter-flies, which includes the peacock and the tortoiseshell. Beetles, craneflies and other small insects are also eaten. During September and October there is fat deposition, and this may weigh as much as 3·8 g.
It is not easy to distinguish between this species and *P. austriacus* (4). The following external features are characteristic of adult specimens of this species: a fore-arm length of less than 39 mm in the male and of under 39·7 mm in the female; a thumb length of more than 6·2 mm in the male and 6·3 mm in the female; a wingspan of less than 265 mm in the male and 270 mm in the female. The head and body length of the male is 39 – 46 mm, and of the female, 37 – 46 mm. The face colour varies from flesh-coloured to light brown, while the overall dorsal colour varies from light buff to grey. The male may weigh up to 10 g and the female up to 13·8 g. The male may live for about 3·7 years and the female for 4·2 years.

4 **Plecotus austriacus** (Grey Long-eared Bat) was only recognized as being a distinct form in Europe in 1959. It has been found in southern Europe and as far north as Poland, Germany, the Netherlands, and now in England. The first specimen in the British Isles was reported in 1964 but it had been collected in 1875 in Hampshire. Subsequently a colony of the common long-eared bat in Dorset was found to contain both species. Other specimens have also been recorded from Dorset and Hampshire. The adults of *P. auritus* are slightly smaller than those of *P. austriacus* which dis-plays sexual dimorphism. Perhaps the best external diagnostic features for distinguishing between the two species are that *P. austriacus* has a fore-arm length of more than 39 mm in the male and 39·7 mm in the female; a thumb length of less than 6·2 mm in the male and 6·3 mm in the female; a wingspan of more than 265 mm in the male and 270 mm in the female. The head and body length of the male is 44 – 51 mm, and of the female 40 – 52 mm. The face colour of the grey long-eared bat is a dark to very dark brown, the overall dorsal colour is grey while the proximal ventral fur is more or less black.
The males and females are rather aggressive when handled, especially if their behaviour is compared with that of *P. auritus* in the same colony.

1 PARTICOLOURED BAT   2 BARBASTELLE
3 COMMON LONG-EARED BAT   4 GREY LONG-EARED BAT

SEROTINE, NOCTULE AND PIPISTRELLE BATS

The bats on this page belong to the order Chiroptera (p. 154) and the family Vespertilionidae (p. 156).

1 **Eptesicus serotinus** (Serotine Bat) is found mainly in the south-east of England with some records from Dorset and Somerset. It hibernates from mid-October to mid-March, usually in buildings. The main colonies live in roofs during the summer, emerging at sunset to feed on beetles, moths, and other insects. The head and body is 74 – 80 mm long, the wingspan 348 – 380 mm and it weighs 20 – 33 g.

2 **Nyctalus noctula** (Noctule) can be found over most of England, particularly in the south-east, and some are present in Wales. Two were recorded in Scotland early in this century.

Males mature sexually in their second year. Females may be mature in their first year, but do not produce young until their second year. Mating occurs in October and living sperm is stored in the reproductive tract of the female (see p. 154). A male will crawl from female to female copulating with each in turn, the females accepting a series of males. The sexes then hibernate together during the winter, from October to March, in buildings, churches, attics, under eaves or in cellars. Hollow trees are used by both summer and winter colonies. In summer the sexes usually segregate, the females often roosting in large breeding colonies. Generally a single young is born in June or July; twins are rare although two records do exist. A female in captivity was observed giving birth. She first arranged herself so that her head was above her body and she was gripping with her thumbs (this is the reverse of the normal resting position). The hind feet were widely separated and the interfemoral membrane formed a pouch to receive the young. Contractions were seen at this stage and birth occurred some 75 minutes later, delivery taking about 20 minutes. Once the head emerged, the mother bent over to lick the young and continued to do so while it struggled to find a nipple. The young bat was very active using thumbs and mouth to cling to and climb the mother. Within a few hours of the birth the umbilical cord was severed and eaten by the female together with the placenta. At birth the young weighs 4·7 – 6·7 g, that is, about 20 per cent of the mother's weight which is 23·8 – 31·2 g after parturition.

At birth the eyes of the young are closed and the body is naked. The thumbs, hind limbs and feet are disproportionately large to enable it to cling to the mother and remain attached during flight. The head is large and 22 deciduous teeth are present in the jaws. At 6 days the eyes open; hair begins to grow at 8 days and the young bat is fully furred at 17 days. Generally growth is rapid, the birth weight being doubled in 14 days. At 6 weeks it is weaned and fully grown, permanent dentition is completed, and it is able to go on foraging flights. Feeding takes place mainly at treetop height but low flights occur during which cockchafers and crickets are taken. The noctule is one of our largest bats; the head and body is 75 - 82 mm long and the wingspan is 353 – 387 mm. It weighs 25 – 39 g, with considerable changes even over quite short periods; one weighed 30 g, 28 g, and 27 g when recaptured on successive evenings.

3 **Nyctalus leisleri** (Leisler's Bat or Lesser Noctule) occurs most frequently in Ireland, but has also been found in Essex, Hertfordshire, Kent, Yorkshire and once in the Shetlands, in buildings and hollow trees. The bats emerge just after sunset to feed on insects. The head and body is up to 63 mm, the wingspan 290 – 320 mm, and the weight 14 – 20 g.

4 **Pipistrellus pipistrellus** (Pipistrelle) is probably ubiquitous on the mainland and is the most abundant bat in Ireland. There are recent records from islands including Soay, Lismore and Scilly. Large colonies roost in houses and churches. One summer colony of some 1000 bats was observed. First a solitary bat would leave about 20 minutes after sunset, then several more 3 – 4 minutes later, the majority following some 10 minutes later, and in about 45 minutes all had left. There was considerable noise and commotion before they emerged and throughout the exodus. This colony caught and ate an estimated 100 kg of flying insects during July and August. An individual was able to catch and consume 0·66 g of insects in 15 minutes.

In summer the females congregate in large numbers to rear their young in the absence of adult males. The young are born in June and July. After the birth the umbilical cord is severed and eaten by the mother within 5 hours. The young, 1·3 – 1·5 g, form as much as 27 per cent of the mother's postpartum weight. At 2 days eye slits appear and the eyes open in about 6 days. The new-born young is very active, and if placed on the ground will crawl and emit ultrasounds with the mouth open, while moving the head from side to side. Hair appears on the rump and centre of the back in 4 – 5 days and by 14 days it is completely furred. The mothers groom the young during the first 7 days. Some young are able to fly at 20 days. As they develop they sometimes form a separate cluster in the colony.

The head and body of this bat is 42 – 52 mm long and the wingspan 200 - 235 mm. Weight varies seasonally being 3 – 4 g in winter and 6·3 – 7·5 g in autumn in preparation for hibernation. Although few records exist of their lifespan, one male was found that had been ringed 11 years before.

1 SEROTINE BAT
3 LEISLER'S BAT

2 NOCTULE BAT
4 PIPISTRELLE BAT

THE BADGER

This animal belongs to the order Carnivora ('flesh-eaters'), characterized by specialized dentition, articulation of the jaw that allows it to open and close but does not permit lateral movement, strong jaw muscles attached to a ridge on the skull, and a well-developed brain. Most carnivores conceal their activities from man. The badger is a member of the family Mustelidae, characterized by musk glands at the base of the tail.

Meles meles (Badger, Brock) is found in England, Ireland and Wales but less frequently in Scotland. It is very widely distributed but often overlooked because of its nocturnal habits. Badgers live in sets which are generally to be found in hilly country, especially if well wooded, 30 – 200 m above sea level and occasionally up to 500 m. Flat country liable to flooding is avoided. Most sets are in sandy soil or in chalk where the drainage is good. They are usually dug into sloping sandy ground in a copse, small wood, hedgerow, or quarry, or into the embankment of a canal, river or railway. The entrance is usually concealed, often by trees, bushes, or thick undergrowth. A set may be used over many years. One of the larger ones, in Somerset, has holes occupying about 0·5 hectare and is several hundred years old. The tunnels are very extensive but there are not necessarily a large number of occupants at any time.

A special set that has several chambers and entrances is used for breeding. It generally takes several years to build and develop. The breeding chamber is an enlarged side tunnel 45 – 60 cm high and 60 – 90 cm in diameter. It is filled with bedding of bracken, hay, leaves, moss, or other material in which the young are almost buried.

It is probable that the boar and sow pair for life. Mating activity appears to be at its peak between February and May just after the birth of the cubs when the sow is on heat. (This period is called postpartum oestrus.) Pairing, preceded by considerable excitement and purring by the boar, lasts from 15 minutes to several hours; pairing of shorter duration does not result in fertilization. The egg reaches the uterus and remains for a long and variable time before becoming implanted in the uterine wall. The final period of development following implantation lasts 40 – 50 days. There may be 1 – 4 young in a litter, but usually there are 3.

Captive badgers of the same species were observed in a zoo in Poland. The boar was excluded from the breeding area for a day after the birth. He remained in the same set, however, but used a different entrance. The new-born cubs were about 90 g, the head and body 12 cm in length, and the tail 3 – 4 cm. They were feeble, blind, toothless, pink-nosed, and covered with sparse white hair. These cubs, a female and a male, weighed respectively 310 g and 320 g at 16 days, 602 g and 640 g at 31 days, 920 g and 980 g at 48 days, 2200 g and 2500 g at 80 days.

Cubs appear above ground when they are about 8 weeks. By 11 weeks, although still suckled, they are very active and will investigate sticks or leaves for some minutes when they emerge. Weaning begins when they are about 12 weeks and takes 2 weeks. Their play during this period is most provoking, and a few weeks later they are seen away from their parents. By 5 months they reach the size of their parents but the coat colouration is not so clearly defined. The female reaches sexual maturity at 12 – 15 months, the male at 12 – 24 months.

During summer and autumn several families may live together and as many as 14 badgers have been seen to emerge from one entrance. The time they emerge each evening varies with the changing light intensity between March and November. Much of the night is spent feeding. They are omnivorous, eating young mammals — especially rabbits — earthworms, land molluscs, numerous insects, and also a large quantity of plant material, including fruit, nuts and grass.

In autumn the sow returns to the breeding set, often the one used previously, removes all the old bedding material, and brings in new plant material until the chamber is almost full. Badgers do not actually hibernate but they all become less active during November and there is a further reduction in activity in December and January. Even during this cold period they rarely remain underground for more than a few days at a time. There is little evidence of territory although it may be confined to the breeding chamber and its surrounding tunnels and entrance. The musk gland at the base of the tail secretes a yellowish, oily liquid. This is emitted either as a result of fear or excitement or to set a scent trail which will enable the badger to find his way home after exploring new areas, and it may be used in marking territory. The chief senses are those of smell and hearing. The badger moves with an ambling trot pausing frequently to listen.

The total length of an adult, including the tail is about 92 cm. The average weight is 12·3 kg for a male and 10·9 kg for a female but there are records of a few animals of 27 kg. Badgers may live for more than 12 years but few data exist.

BADGER

THE FOX

Vulpes vulpes (Fox) is a member of the order Carnivora (p. 164) and the Canidae, or dog family, of which it is the only wild representative in this country. Characterized by long muzzle and ears, large brain case, and non-retractile claws, it is a swift-moving predator. It is probably nearly ubiquitous in England, Wales, Scotland and Ireland, where considerable numbers remain as it has long been preserved for hunting. It is absent from most of the smaller islands where its history is complicated by introduction and extermination, the only recent record being for Scalpay in Skye.

The fox can be found from mountainous to lowland areas, frequently on scrubland and sea-cliffs. It is even found on the edge of towns, including London where it has been reported within 17 km of the centre. Foxes living close to a town will range widely in search of food as well as taking food from dustbins.

For its den it shows some preference for sandholes but will often use another animal's burrow — for example, a large rabbit burrow or a badger set — or a peat hole or rock cairn. It brings food there so that the neighbourhood of the den is often littered with bird and animal remains. When in its den the fox is said to have 'gone to earth'.

The breeding den must be dry and in reach of an adequate food supply. It is selected by the vixen after the dog has searched for a suitable site. The dog mates only once with his chosen vixen, usually between January and March depending on the region. The dog is fertile for a considerable period before and after the three days or so during which the vixen is on heat. Mating is prolonged, the pair becoming knotted together rather like the domestic dog. The gestation period lasts for 51 – 63 days. Towards the end of this time the vixen tends to remain almost entirely below ground, and the dog will carry food to her.

The cubs are usually born between March and May and the dog will spend considerable time with them. If the breeding place is disturbed, the cubs are moved to a safer place. There are usually 4 – 5 young in a litter, but up to 8 have been recorded. There is evidence of resorption of embryos in some vixens resulting in an overall decrease in the number of embryos during pregnancy.

When newly born the cub has a blue-grey, woolly coat with a white tip to the tail. At 3 days a change in the colour begins; brown appears first on the head and this gradually spreads until at 3 weeks it covers the head and shoulders. By 4 weeks the cub is a dark fawn-brown. Between July and September a fuller moult occurs to the adult pelage, and the muzzle elongates.

At birth the cub is blind; the eyes open about 8 days later. It is without teeth until the milk ones erupt in 3 – 4 weeks. There is a gradual change from a milk to a meat diet which begins when the vixen gives the cub flesh to suck or food which she has regurgitated. Most cubs are weaned by 30 – 40 days, and then live on flesh, mainly voles. At first the vixen will bring home disabled prey for the cubs to practise hunting and killing.

Growth is fairly rapid. At birth a cub weighs 100 – 110 g, at 3 days 235 g, at 14 days 800 g, and after 28 days 1 kg. By 13 weeks a weight of about 3 kg is reached, and by 6 – 7 months 5 kg. The cub is regarded as an adult when it is about 8 months.

The adults live largely overground from April onwards, the dog ranging over a wider area than the vixen. Tagging experiments revealed that adults had not moved more than 5·5 km from the site of marking. The cubs also live mainly overground from July on, and between August and November when 6 – 9 months of age they make long migratory movements. Distances of up to 30 km are not uncommon amongst cubs of this age, and one of 58 km has been recorded.

The adults feed chiefly upon rabbits, young hares, birds, voles, mice, rats, sheep, deer, and some insects. Each animal needs daily about 400 g of food, or the equivalent of one young rabbit. The rabbit formed a major part of the fox's diet before the advent of myxomatosis, and does again since the disappearance of the disease. Foxes will also eat birds, however, for it was found that many foxes died after taking pigeons and other birds that had been eating seeds treated with dieldrin or other similar organochlorine compounds.

The head and body of the adult fox is 65 cm long, and the tail 40 cm. The dog fox can weigh as much as 9·2 kg, the vixen 6·9 kg and when pregnant 9·5 kg. A few survive to 4 years and rare examples have been known to survive to 6 years.

FOX and CUBS

POLECAT, FERRET AND MINK

The animals on this page belong to the order Carnivora (p. 164) and the family Mustelidae (p. 164).

1 Mustela putorius (Polecat) is now found throughout most of Wales where there has been a recent spread, the only exception being in the extreme south. The last of those living in Scotland were trapped in 1907 and a few remained in parts of England until the 1930s.

In Wales it is found in a wide variety of habitats, from dunes to mountainous areas up to 450 m. It is most common, however, in thickets and woods where its den is made amongst rocks, tree roots or in rabbit burrows after it has eaten the occupants. In very cold weather it may move near or even into farm buildings or outhouses to find shelter where food, such as rats and mice, is also available. Although not a good climber the polecat can swim well.

Little is known of the social habits of this animal but it will mark its own territory with urine and secretion from the anal gland. It spends the day in the den and comes out at night to hunt. It locates most mammalian prey by sound and scent, catches it by the head or neck, and then shakes it violently. Its prey includes young rabbits and hares, water voles, short-tailed voles, woodmice, hedgehogs, birds, the common frog and the lizard, as well as invertebrates.

Although little precise information about its reproductive habits is available, it seems probable that there is one litter per year. The male, or hob, and the female, or jill, mate in April. The gestation period lasts 40 – 60 days, and the 4 – 5 young are born in June. The newborn kitten has a thin covering of white hair which is replaced at 20 days with a dark coat. When it is 50 days old the typical colouring of the adult is apparent. In the first few days of life it is able to cry and by 35 days can hiss. The eyes open at 6 weeks and by 8 weeks the kitten is very active. It will drag and shake objects, snap its jaws, and chase around in the proximity of the den. At 9·5 weeks it can jump on an opponent and by 14 weeks is very rough.

The average length of an adult male is 52 cm from nose to tip of tail and the weight is 957 g, while a female is 46 cm and 623 g. The winter coat appears much lighter than the summer coat for its fur is longer, denser, and whiter. The polecat was extensively trapped for its fur in the early 19th century but rarely now.

2 Mustela putorius furo (Ferret), a close relative to the polecat, is essentially a domesticated animal that has been escaping over many years. The majority are albino this character having been selected not only for the fur colour but also because albinos are more docile and easily domesticated. However, many escapees have interbred with polecats and produced colour variations.

The ferret is found in Yorkshire, Lancashire, Somerset, and the Isle of Man, and has recently been reported in Aberdeen and the Isle of Mull.

It will live wherever there is cover and food but prefers scrubland, copses, and wood in lowland areas, although it will go up to 600 m. The ferret preys upon young rabbits and hares, birds, amphibia, reptiles, fish and eggs, and will often remain fairly close to habitation where food can be obtained. It does not become sexually mature in its first year. There is one breeding season of 6 months each year, during which two litters are usually produced. The pairing of the hob and the jill may last up to 3 hours. There is no delay in implantation and the gestation period lasts 42 days. There are 6 – 8 young in a litter and each weighs 9 – 10 g at birth, while the mother weighs about 700 g. The kitten is practically naked at birth. After about 4 weeks the eyes open but it will crawl about and eat solid food earlier than this. By 6 – 8 weeks it is weaned and very active.

3 Mustela vison (American Mink, Mink) is a North American species from which all minks in the British Isles are descended. Fur farms were established in 1929 and animals have been escaping and living ferally since then. It was first found to be breeding in the wild in 1956 in the upper reaches of the River Teign in Devon. Now it is established in the south and north of England, in south Wales, and in the southern half of Scotland. More elusive and agile than the weasel, it is a very successful hunter. Its prey includes geese, ducks, coots, moorhens and their eggs as well as poultry and game birds.

Largely aquatic, it spends much time in rivers and lives in close proximity to water although it will also move some way from it and can cover quite considerable distances. One mink that has been recorded travelled some 4·8 km in 2 days and at least 80 km along a river in 3 years. Although difficult to find, minks have been seen at all times of the day but are probably most active in poor light.

There is a well-defined breeding season in captivity, the female being receptive to the male for a few days only between late February and early April. Pairing lasts for 30 – 40 minutes. A short delay, of variable length depending upon the time of mating, occurs before implantation. The gestation period lasts for 42 – 53 days, the young being born about mid-May.

The head and body is 302 – 430 mm and the tail 127 – 229 mm; it weighs 565 – 1020 g although the male in captivity may be much heavier.

1 POLECAT
2 FERRET
3 AMERICAN MINK

SMALL CARNIVORES

The animals on this page belong to the order Carnivora (p. 164) and the family Mustelidae (p. 164).

1 **Martes martes** (Pine Marten) is now found only in the mountainous parts of Scotland, Wales, England and southern Ireland. Previously more widespread, it has been persecuted for its fur. Predominantly an inhabitant of coniferous woods, it will, however, inhabit open, rocky ground where it may have its den in a cairn or rock crevice.

Small prey is preferred and rodents and small birds form the bulk of its food throughout the year. Young hares and rabbits as well as fish, butterflies, moths, beetles, and fruit are eaten. The carrion of deer, fish, and birds is also taken.

Oestrus occurs in July and August after the young are weaned and at this time the female sets scent and urinates frequently. Copulation has been seen during the day, at dusk, and after dark. It lasts about an hour during which time the dog drags the bitch by the scruff of her neck whilst purring and growling. There is then a delay in implantation which takes place in mid-January; the final period of development lasts about 55 days. Usually 3 young are born in March or April and are weaned in 6 – 7 weeks. Only one litter is produced per year. Sexual maturity is reached at 2 years.

The head and body of the male is 480 – 530 mm and of the female 400 – 450 mm, while the tail is 250 – 280 mm and 230 – 260 mm respectively. Their weight varies between 900 and 1500 g depending upon the season but rises to a peak in June.

2 **Mustela nivalis** (Weasel) is probably ubiquitous on the mainland at all altitudes but is absent from Ireland and most of the small islands with the exception of Skye, Anglesey, and Wight.

It is widespread in most habitats from lowland farms and woods to moorland and mountain. In a forest plantation its territory may cover about 2·2 hectare. During the day it may be seen standing vertically on its hind limbs looking at the surrounding area, but most of the hunting is done by night. When alarmed it gives a guttural hiss and if provoked gives a short, screaming bark. The young have a shrill scream.

The weasel can climb and swim well and is thus able to take a wide range of prey. Mice, rats, shrews, voles, squirrels, moles, rabbits, hares, perching birds and eggs are eaten. Frogs, toads, lizards, crayfish and insects are also eaten but the bulk of its food consists of small rodents. It takes about one-third of its own weight in food each day. If hungry it will resort to eating carrion. It pounces on small prey and bites them at the back of the neck with the needle-like canine teeth which enter the skull and damage the brain.

The dog is ready to breed in March and the first pregnant bitches are found in this month. There is no delay in implantation. The gestation period lasts about 35 days. There is a peak in births during April and May. Some females have a second litter in August. There are usually 3 – 8 young in a litter. At 1 day a kitten weighs 2 – 4 g and is 50 mm long. The young can suck and chew pieces

of mouse at 2 weeks and a few days later eat solid food. Growth continues fairly rapidly and by 30 days the young male is some 142 mm and 67 g, and the young female is about 137 mm and 48 g. By July the young of an early litter reach adult size, and the females may produce a late litter in the same season.

The head and body of the dog is 175 – 220 mm, and of the bitch 150 – 190 mm, the tail being 40 – 75 mm and 40 – 65 mm respectively. The dog weighs 70 – 170 g, and the bitch 35 – 90 g. Although different in size both sexes are similar in appearance. A weasel may live for about 6 years.

3 **Mustela erminea** (Stoat, Ermine) is probably present throughout the British Isles except on small islands although it has been introduced into the Shetlands. It lives in agricultural areas, marsh, woodland, moorland, and mountain. The den may be in a hollow tree, a rock crannie, or an old rabbit burrow, and the territory covers about 20 hectare. It moves with a long bounding gait and can swim well. If alarmed it gives a bark or spitting rattle.

The stoat regularly preys upon animals several times its own weight, and has been seen to attack and kill a mountain hare. It usually kills rabbits by biting them repeatedly with the canine teeth at the back of the neck. Rodents including rats and mice are eaten as well as perching birds, pigeons, and game birds. It eats about one-third of its own weight each day, and drinks frequently.

Most young are born in March or early April. Mating occurs some time during the 5 weeks following parturition, while the female is still lactating. Implantation is delayed until the following spring when development to full term takes a further 21 – 28 days. There are 6 – 13 young in a litter but the usual number is 9. At 7 days the kitten has a fine covering of silvery hair and by 14 days there is a faint darkening on the back. By 21 days both sexes show a conspicuous mane. When 35 days old the kitten is able to eat mouse flesh. The eyes open between 36 and 41 days, the female's before the male's. The kittens begin to play at 7 – 8 weeks and can kill live prey at 11 – 12 weeks. As they grow up they go hunting with the parents.

The coat undergoes two changes: the spring moult is slow and proceeds from the back to the belly; the autumn one, in contrast, is rapid and proceeds in the opposite direction. In the southern part of the country the colour remains the same throughout the year but the fur is denser in winter. In the north the pelage becomes entirely white except for the black tip of the tail. There are, however, records from a narrow zone in Yorkshire where both colours may occur in winter. The head and body of the adult dog is 275 – 312 mm and of the bitch is 242 – 292 mm, the tail 95 – 127 mm and 95 – 140 mm respectively. The dog weighs 200 – 445 g and the bitch 140 – 280 g. Those in Ireland tend to be smaller.

1 PINE MARTEN   2 WEASEL
3 STOAT, summer coat   3A STOAT, winter coat   3B STOAT, kitten

THE WILD CAT

Felis sylvestris (Wild Cat) is a member of the order Carnivora (p. 164), and the family Felidae. This family feeds almost exclusively on vertebrate prey and is the most specialized one of the order. The jaws, operated by powerful muscles, are short and bear large canine and carnassial (shearing) teeth. The feet have become adapted as additional weapons of attack with their sharp, retractile claws. Sight is very important while smell is subordinate to it, in contrast with other carnivores.

It is present in Scotland mainly in the Highlands although there are two recent records from Lanarkshire and Stirlingshire south of the Highland boundary. It is widely distributed throughout the Highlands especially in woods and moors but is scarce in the north and west. On the higher moors of Inverness-shire and Aberdeenshire it is more numerous but is uncommon in Perthshire. During the early 1960s there was a marked and sudden increase in the population in some of the Angus hills, and also in parts of Moray and Nairn. Before the end of the 19th century it was also found in the lowlands of Scotland, and in 1837 it was not uncommon in England and Wales. The predatory habits of the wild cat led to its extermination in these areas but several large landowners gave it protection in the Highlands of Scotland, where it remains today. There is considerable doubt about the presence of the wild cat in Ireland and recently it has been suggested that the domestic cat was first taken there in the Bronze Age, perhaps in connection with religious beliefs.

Generally considered to be one of the wildest and most untameable of mammals, it is a solitary animal. It avoids man and lives in hollow trees in isolated woods or in holes in rocky places, usually high above sea level. The wild cat tends to have only one mate, and each pair appears to have their own territory. From this they will travel far, especially in winter or during the breeding season, to find prey. Man is most likely to encounter it on the open deer moors which it frequents in search of food. The home range of some 60 – 70 hectare is defended by the male. A nocturnal animal, it emerges at night in search of prey which includes hares, rabbits, squirrels, lambs, fawns and other small mammals. Game birds are often taken and rarely frogs and fish. Prey, stalked stealthily, is caught by a sudden pounce. Although the pure-bred wild cat is said to breed only once a year, the Scottish wild cat breeds twice, and even three times; this may be due to inbreeding with feral domestic cats. The tom is mature sexually when 10 months of age and the puss when less than 12 months. They can breed in their first year. Sexual activity in the male lasts from the end of December to the end of June. The puss is in oestrus for 5 – 6 days during the first half of March and the kittens are born in April or May. A second oestrus may occur towards the end of lactation, at the end of May or beginning of June, and may result in another litter being produced in August. Occasionally there is a third breeding season in the autumn, the young being born in winter.

The puss generally makes a nest in a hole or rock crevice away from the tom as he may kill the newborn young. The gestation period lasts for 63 – 69 days; litters of up to 8 young have been reported but usually there are 4 – 5 kittens. The ground colour of the coat is light with greyish-brown tabby markings. When 4 – 5 weeks of age the kitten will leave the nest but does not hunt with the mother until 10 – 12 weeks of age. By 6 weeks it weighs about 0·5 kg, and by 8 months 2·7 kg. It is weaned when about 4 months of age and leaves the mother when 5 months old. The mortality is high among kittens, many dying when 2 – 4 months. Both the tom and the puss have been known to devour young under various circumstances.

The head and body of the male is 572 – 653 mm long and of the female 495 – 571 mm; the tail is 286 – 370 mm and 257 – 320 mm respectively. The male is 5 – 6·8 kg in weight and the female 3·8 – 4·5 kg; only rarely are heavier specimens found. The sounds it produces are, like those of the domestic cat, a miaow, a purr when pleased, and a growl when angry.

WILD CAT

THE OTTER

Lutra lutra (Otter). This semi-aquatic, nocturnal animal is a member of the order Carnivora (p. 164) and the family Mustelidae (p. 164). Although not often seen, it is still found on the mainland and is common in Ireland. There has been a considerable reduction in the population in many areas but the distribution has shown little alteration. Some possible causes for the decrease may be pollution of the rivers, particularly severe weather, competition with escaped mink, and the increased popularity of fishing.

Slow-moving rivers with plenty of ground cover are preferred although mountain streams are also inhabited. Otters spend the day lying up in their resting place, or holt, usually a disused burrow, drain or hollow tree. Throughout the year, in some areas of Scotland, they use couches which they have made beneath dense shrubby growth from sedge, rhododendron, rushes or rough herbage, and to which they add material in the autumn. Typically, the otter hunts for a few nights along a stretch of river and then moves to a new feeding ground. Those living by salmon or trout streams often follow the fish up to the spawning beds and spend the winter in the head waters away from the flooded part. In the spring they follow the young fish downstream and then spend some time around the estuary and along the sea-shore.

In parts of western Scotland and the Hebrides some otters have become entirely marine. Islands 8 km offshore have been colonized where the otters feed exclusively on sea fish. They use rock crannies or caves as holts and on occasional journeys inland they lie up in reed beds.

They choose a site such as a prominent boulder or an old tree trunk to leave their droppings or spraints. All the otters in a district will regularly visit these places, which may serve for communication between individuals by means of scent. This is secreted by two small glands at the base of the tail. It is this scent, or drag, often very strong, that is picked up by hounds during otter hunts.

In winter, particularly in conditions of snow, their tracks, or seals can be followed. Otters run a few steps and then slide for some 3 – 6 m on their stomach. These slides may be seen with seals at each end. This animal is well adapted for its life in the water, most obviously perhaps in its streamlined shape. The feet have widely separated toes with strong interdigital webs for swimming. The tail, rudder, or pole, is more than half as long as the head and body together and is of great value in the water. The rather small eyes are on top of the head enabling the otter to see when almost submerged. The small rounded ears and valvular nostrils can both be closed at will as the otter dives below the surface. It can remain submerged for quite long periods as the lungs have a large air capacity. The vibrissae, or whiskers, set on well-developed pads on the snout and lips, have a complex nervous supply and form very sensitive tactile organs. The undercoat of close fur is waterproof and traps air bubbles to form an insulating layer. When the otter emerges from water, the fur is bunched into spiky tufts but soon dries after shaking. Much time is spent rubbing and grooming.

When hunting it emits piercing whistles that can be heard over long distances. The prey is taken to a particular site where it is bitten and chewed before being swallowed. In freshwater eels, fish, frogs, toads, moorhens and small mammals are taken. Otters living by the sea catch crabs and other crustacea as well as flounders, cod, wrasse, and flatfish. Evidence of the type of food eaten can be found from examination of the faeces or spraints. Carrion is also sometimes eaten. A young, growing otter, three-quarters the size of an adult, can eat 5 – 9 kg of food in a week.

Sexual maturity is reached in 2 years and during breeding there are indications of territorial behaviour. The dog can be very aggressive at this time and fierce fighting will occur if two dogs follow the same bitch. Mating takes place in water and is prolonged, lasting 1 – 1·5 hours, during which time the pair continue to whistle and chatter. If a female has produced young, then mating occurs in the post-partum oestrus. There is a delay in implantation of the blastocyst, and the final period of development lasts 61 – 63 days.

The only time an otter remains in one place is when the bitch has young cubs. A favourite site for the breeding holt or den is up a small side stream, away from the main river and floods, in an old burrow or drain. The nest is formed of reeds or dried grass and lined with moss, fur, or wool. There are usually 2 – 3 young in each litter although as many as 5 may be produced. Most of the care and upbringing of the cubs is left to the bitch who remains with them for long periods in the first few weeks. The dog usually remains downstream in the vicinity.

At birth the cubs are blind and toothless and have a covering of fine, dark, silky hair which becomes rougher and lighter in colour as it grows longer. At about 35 days the cubs are able to see; they seem unable to focus upon objects initially and, when they begin to use their eyes, are short-sighted. They remain in the den for about 8 weeks by which time they are weaned. Shortly afterwards they explore the area near the den, climbing almost anything within reach, but avoiding water. This last perhaps somewhat surprising, feature of their behaviour serves them in good stead as the under-fur is not sufficiently thick to be waterproof until they are about 3 months old. This is the usual age at which they learn to swim. At first the bitch may have to coax the cubs into the water and sometimes even force them, and once they are able to swim, they have to learn to catch fish.

The total length of the male is 96 – 136 cm and of the female 94 – 112 cm, while they weigh 5·5 – 16·8 kg and 6·4 – 12·3 kg respectively; rare records of 23 kg.

OTTER and CUBS

THE COMMON SEAL

Seal and walrus belong to the order Pinnipedia, animals well-adapted for aquatic life but which must surface at intervals to breathe and to return to land or ice to breed. The body is smooth, streamlined, and insulated by a thick layer of fat, or blubber. The limbs have developed into flippers for use in water, and the hind ones are turned backwards making them extremely efficient swimmers. Their progress on land is, however, relatively poor. They feed exclusively on animals but are more sociable than land carnivores. The seals found around the British Isles are true or 'earless' ones that belong to the family Phocidae.

Phoca vitulina (Common or Harbour Seal) is found in considerable numbers, particularly during the breeding season, on the coasts of East Anglia and the Wash in England, on the Shetlands, Orkneys and Hebrides and the east and west coasts of Scotland, and on the north-east coast of Ireland. At other seasons of the year it is somewhat more widely distributed around the coasts. Generally it lives on sand or mud banks in harbours or estuaries but in the Shetlands has become adapted to exposed rocks.

Resting sites must provide shelter and ready access to food and have a gradual, underwater slope. In such an area this seal will remain throughout the year, except during storms when it stays at sea for days or weeks. Two or more seals will haul out on to sand or rocks on the ebbing tide and take to the water again on the flood tide. A colony will consist of seals of all ages and both sexes. Generally one bull will occupy the highest point and watch while the colony is at rest. No other territorial behaviour appears to be exhibited by bulls, although a mother and pup may occupy a particular small area of the beach. There is little interaction between cows.

The bull mates with only one female. He is mature sexually when 6 – 7 years and the cow at 5 – 6 years. Mating occurs in the Shetlands between early September and early October and on the coast of East Anglia in July and August. It takes place some 4 weeks after lactation has ceased and when the faded coat has completely moulted. Pairs may be seen, early in September in the Shetlands, rolling over one another in the water, writhing and twisting in almost continual contact whilst yelping and snarling and slapping their flippers on the water until it foams. This behaviour can end in coition but bubble-blowing may occur before it does so. The pair either dive or sink to the bottom and as they float up blow a fountain of bubbles. After a variable period of these activities the male will mount the back of the female. He grasps her in the axillary position with his fore-flippers, often while mouthing her head. The pair sink until close to the bottom in water 4 – 8 m deep. They remain in coition, one on top of the other at a slight angle so that their hind ends can twist until in contact. In May adults and sub-adults will roll and play in 'pairs', but it is not known whether this leads to coition.

There is a period of 2 – 3 months delay before implantation of the embryo on to the uterine wall takes place in November or in December. The gestation period continues and the young are born between May and July, depending upon the region. Birth usually takes place on low tidal rocks or on sandbanks. The single pup is initially very slim, 75 – 93 cm long, 9 – 11 kg in weight. The embryonic coat of long white hair is shed either *in utero* or soon after birth, leaving a coat of short hair. The pup can swim immediately after birth but spends much time within 20 m of the shore floating limply at the surface close to the mother, often in physical contact. At 2 days the pup can swim well and, when below the surface, is taken between the fore-flippers of the mother while learning to dive. Mother and pup play together and the pup often rides on its mother's back. The pup can suckle for up to 6 minutes below water in a calm sea while the mother floats in a prone or vertical position or rests on a sea-washed rock. At first the pup comes ashore only with extreme difficulty as the hind limbs lack the muscular power necessary for movement on land. When about 10 days old it is able to come ashore for longer periods, and subsequently the mother and pup do so regularly. At 3 weeks the pup has filled out and become streamlined. It is now able to suckle on land and remains with its mother until lactation is completed, 4 – 6 weeks after birth. The rich milk, containing 9 per cent protein and 45 per cent fat, ensures rapid growth and reduces the period of helplessness. After weaning the pup eats shrimps and prawns. Adults eat flatfish such as dab, flounder and plaice as well as whiting, cod, eel and gobies. Crustacea and molluscs are also eaten. Digestion is rapid, food traces being lost from the stomach within 3 hours. They normally feed in daylight at high tides. The common seal is much more aquatic than the grey seal, although it generally stays in shallow waters. It is a good swimmer, can leap clear of the water, and can dive and remain submerged for as long as 28 minutes but seldom stays so long. The maximum recorded dive is one of 91 m, but such a deep dive would be of very short duration.

The adult male is 153 – 198 cm from head to tip of flipper and the female 137 – 168 cm; they weigh 200 – 253 kg and 100 – 150 kg respectively. In captivity a male has survived for 18 years and a female for 14 years.

This seal may be hunted by the killer whale but its chief enemy is man, who, on occasion, has killed an entire generation of pups for their pelts. The kitchen middens of Bronze Age man have revealed that he ate seal meat.

COMMON SEAL and PUP

THE GREY SEAL

Halichoerus grypus (Grey Seal) is a member of the order Pinnipedia (p. 176), and the family Phocidae (p. 176). It is widespread round the British Isles, and indeed 65 per cent of the world population of some 52,500 is found along our coasts. The large breeding colonies on North Rona and Sula Sgeir, about 80 km off the coast of Scotland, account for about half of the world population. These seals are also found on the islands of Orkney, Hebrides, Shetland, Farne, Man, and Scilly, on the coast of Devon, Cornwall, and Dorset, and along the Welsh coast and on Ramsey Island. Exposed, rocky coasts are preferred both for hauling out on to and for breeding.

Grey seals assemble during the autumn, the cows on off-shore rocks near the sea to give birth to their young, and the bulls on rocks and islets to establish territories, for mating takes place some 14 days after parturition. During this period the adults starve. This species chooses to breed under crowded conditions as they will inhabit one island while another close by, and apparently equally suitable, is unoccupied. Pups are born between September and December around the British Isles; there is a short peak period peculiar to the colony and the region. The birth of a pup is very rapid, lasting from one to several seconds. The cow breaks the umbilical cord by a sharp movement of her hindquarters. The placenta is delivered a few minutes later and is usually eaten by seabirds. Many cows stay with their newborn calf to nuzzle it, others leave immediately but return a few hours later to feed it.

The cow will come back to suckle the young several times during the first 24 hours, but subsequent visits are less frequent. Suckling will last continuously for some 6 minutes followed by intermittent feeding for up to 16 minutes. The milk is rich, about 52 per cent fat and 11 per cent protein, and growth is rapid between birth and 17 days when suckling usually ceases. At birth the pup weighs about 14 kg. One weighed 19 kg at 3 days and 42 kg at 18 days, the increase being about 1·5 kg per day, and mainly in girth rather than length, for there is a considerable deposition of fat.

The pups remain on the nursery beach and, if undisturbed, show no antagonism but lie and sleep. They may provoke bites from other pups or cows by moving. It seems likely that hungry pups will call their mothers from the sea, those nearest to the shore being more successful in attracting attention. Survival is greatest in pups which have achieved a good rate of growth initially, but mortality is generally quite high.

The newly born pup is covered with a fairly long, silky white coat, with some dark fur on the face and the flippers. At 3 weeks it begins to moult and in 1 – 2 weeks the first coat is shed and the adult pattern appears. This period of moulting coincides with the cessation of suckling when the cow leaves the pup. These events result in considerable loss of weight. The pup remains on the beach until the moult is complete and then goes to sea where it has to learn to swim and to feed itself. The permanent teeth usually begin to erupt about 4 days after birth, the milk teeth having developed and been shed *in utero*.

The dispersal of pups takes place when they go to sea, some remaining close to the nursery beaches but others travelling considerable distances. One pup marked in the Farne Islands was recovered 14 days later some 640 km away off the Norwegian coast.

While the cows are occupied with the birth and feeding of their young, the bulls gather in large groups on rocks and islets in an area adjacent or close to the nursery beach, to establish and defend their territories. If one bull were to enter the territory of another the occupant would charge, open his mouth, and perhaps snarl. Real battles rarely ensue as the trespasser normally retreats. When a cow comes into oestrus some 14 days after parturition she does not oppose the advances of a bull. Mating usually takes place below the surface in shallow waters but it can occur in surf or even on land. Copulation lasts for 15–30 minutes, the pair periodically coming to the surface for air.

After fertilization the ovum develops to the blastocyst stage when growth stops and it remains in the uterine horn. There is a delay of about 100 days before implantation, and most of this time the cow spends entirely at sea, feeding and recovering from the effects of parturition and its accompanying period of starvation. The cows then come ashore to moult — as the bulls do some 3 months later. The blastocyst then continues its development; the remaining part of the gestation period lasts about 250 days. Thus there is a period of some 14 days each year when a mature cow is not pregnant.

The grey seal exhibits sexual dimorphism. The bull has much blubber on the neck and shoulders, the skin is thrown into folds, the profile is rounded, and the muzzle heavier and broader than the cow's. She has a sleek neck and a straight profile. Maturation in the male takes place gradually and is complete by 9 – 10 years of age; at 10 or more years he will begin to hold territory. The length from the snout to the tip of the hind flippers is 216 – 254 cm in the male and 185 – 220 cm in the female. A male 254 cm long weighed 131 kg and a female 203 cm was 90 kg. The age of this seal can be ascertained from the canine teeth, and one female, 186 cm long, was found to be 46 years of age. Only a few are likely to reach such an age.

GREY SEAL

1 bull 1A cow 1B pup

SEALS AND WALRUS

All these animals belong to the order Pinnipedia (p. 176); the seals belong to the family Phocidae (p. 176), and the walrus is the only member of the family Odobenidae. The walrus is able to turn its hind limbs forward when moving on land, but has no external ears.

1 **Phoca groenlandica** (Harp Seal) is found in the open seas of the Arctic region of the Atlantic ocean. Harp seals spend the summer in the north and then migrate southwards in spring to gather in immense herds on the ice to breed. Stragglers occurred occasionally in the last century in the Shetland Islands, Scotland, and as far south as the River Teign in Devon.

Males reach sexual maturity at 8 years, females at 6 years. Mating occurs about 14 days after parturition and this is the only time the male is seen on the ice. Each male mates with only one female. There is a delay of 11 weeks before the blastocyst is implanted. The pups, 60 – 76 cm long and 4·5 kg in weight, are born in February and March. The coat is of stiff white fur which is shed in 3 – 4 weeks, when the pup takes to the sea to feed on small crustacea. The adult male reaches a length of some 170 cm and a weight of 180 kg but the female is slightly smaller.

2 **Phoca (Pusa) hispida** (Ringed Seal) is one of the commonest seals of the circumpolar Arctic region, found in open water close to ice or land but not usually in the open sea. Rare stragglers appear on British coasts and have been recorded from Scotland, St. Kilda, Norfolk, and Galway Bay, Ireland where one caught in 1895 lived for some time in Dublin Zoo.

Sexual maturity is reached at 7 – 8 years. The adults mate within 2 weeks of parturition and there is a delay in implantation of 3 – 4 months. The young are born on land-fast ice in March and May and are left either in a lair beneath the snow or in a natural hollow with a breathing hole. The pup at birth is 60 cm long and 4·5 kg in weight. Interestingly, this seal has a long period of parental care, and lactation may last for 2 months. The adults are about 140 cm long and weigh some 90 kg, both sexes being in the same size range. One male of 43 years has been found.

This seal is important in the economy of the Eskimo who eat the flesh, liver and intestines, burn the blubber in lamps, and use ths skin for clothing and for covering their kayaks.

3 **Cystophora cristata** (Hooded Seal) is usually associated with drifting ice floes in the deep waters of the Atlantic ocean. Rare stragglers have been recorded, mainly in the 19th century, from Ireland, Scotland, and England. The world population is about half a million.

Until sexual maturity is reached at 3 – 4 years, it is solitary but then groups are formed of separate sexes. Family groups gather when the pups are born in March and April. Immediately after lactation the adults mate,

12 – 14 days after the birth of the young. They then return to the sea but assemble again in June and July to moult. The pup at birth has a beautiful coat, silvery grey on the back, darker on the face with a sharp demarcation from the creamy ventral surface. As the pup remains on the ice floe for 4 weeks after birth, it often falls prey to the hunter who wants the skin.

The male has a sac beneath the skin above the muzzle, connected with the nasal cavity, that can be inflated to form a high cushion on the top of the head, the 'hood'; it also possesses a nasal sac which can be blown out into a balloon-like structure. The function of these curious features is not yet understood. Adults reach a length of up to 245 cm and a weight of some 400 kg.

4 **Odobenus rosmarus** (Walrus) is found in shallow water off Arctic coasts where it remains close to land. In 1456 William Caxton, the printer, recorded a walrus taken in the Thames. More recently, 1815 – 1954, there have been records of 21 seen or killed around the British Isles, the majority being from Scotland, particularly the Shetland Islands where they occurred more frequently in the 19th century. The last one in Shetland waters was present from September to November, 1926; it was about 3·6 m in length and its canine teeth were about 30 cm long.

Sexual maturity is reached at 5 – 6 years in the female and a year or so later in the male. Gestation lasts for about a year and pregnancy occurs probably every second or third year. The young are born in April and May usually on ice and are said to remain with the mother for 2 or 3 years. Lactation lasts at least 16 months. The upper canine teeth or tusks grow throughout life in both sexes and may reach a length of 100 cm in the male and 60 cm in the female. These teeth are used to dig clams, whelks, and other molluscs from the sea bed. A single tusk from an old male may weigh as much as 5·35 kg, and it is for this ivory that the walrus has so often been hunted.

5 **Erignathus barbatus** (Bearded Seal) is a circumpolar species which rarely enters more temperate regions. It prefers shallow water near the coast, gravel beaches, and ice floes not far from land. Stragglers have been reported from Scotland and Norfolk. Adult males and females are of similar size, about 230 cm long and 220 – 270 kg in weight. Sexual maturity is reached at 6 – 7 years. The young are born in May and June and mating occurs two weeks after parturition. The great profusion of whiskers, or vibrissae, gives this seal its name.

1 HARP SEAL 2 RINGED SEAL

3 HOODED SEAL

4 WALRUS 5 BEARDED SEAL

THE COMMON DOLPHIN AND PORPOISE

Members of the order Cetacea are highly adapted for aquatic life. The fish-shaped body is hairless and has an insulating layer of blubber; the forelimbs have developed into flippers and the hind ones are lost; and the trunk expands posteriorly into two horizontal fleshy lobes or flukes. These carnivores can feed under water as there is no communication between the mouth and lungs, the respiratory outlet being the blowhole(s). The vascular and respiratory systems are well-adapted for diving. There are two sub-orders: the Odontoceti, characterized by teeth, a single blow-hole, and long, slender flippers; and the Mysticeti, whalebone whales, which lack teeth. The common dolphin (1) belongs to the family Delphinidae which are typically beaked; the common porpoise (2) is of the family Phocaenidae and lacks the beak; both families belong to the sub-order Odontoceti.

1 **Delphinus delphis** (Common Dolphin) is an inhabitant of the warmer parts of the Atlantic and the Mediterranean. It comes as far north as the British Isles where it is a common visitor in the English Channel and off the coast of Ireland. On rare occasions it has been recorded from as far north as the Shetland Islands. Sometimes considerable numbers are stranded in small estuaries when the tide has cut off their retreat; this happens most often in early spring or late summer. Although usually oceanic, there are records of movements up rivers; in 1947 a school of 20 was reported in the River Nene, Cambridgeshire, and others have been seen in the River Thames even as far up as Chiswick. These inshore migrations may be correlated with the movement of prey, as a school will pursue shoals of herring, sardines, mackerel and pilchards and devour immense quantities. Fish is the chief food although squids and crustacea are also eaten.

A school may be of some 200 individuals and it is a beautiful sight to see them leaping clear of the surface almost synchronously. Indeed a dolphin of 150 – 200 kg can leap 3 m into the air. Amongst the swiftest of the cetaceans, it can maintain a speed of over 10 m per second for several hours, as is well known from its habit of accompanying boats. It is thought that a dolphin reaches sexual maturity at 3 years and may produce one young each year. Mating probably occurs in spring and summer and the gestation period lasts 11 months. The young at birth is 75 – 90 cm long. A single female calf 106 cm long was stranded on the Cardiganshire coast one July. A length of up to 2·4 m can be reached by the adult and one of 2·1 m weighed about 110 kg.

2 **Phocoena phocoena** (Common Porpoise) has been known in the Mediterranean since Classical times. It is the smallest and most common cetacean in European waters and is widely distributed in the North Atlantic, the North Sea, and the English Channel. It occurs all around the British Isles, usually in summer or early autumn when it appears in coastal areas and may travel many miles up river. In summer it approaches the coast, and in the Shetland Islands it will penetrate narrow, shallow inlets. It is common all around Scotland, particularly when small fish are plentiful, and large schools will stay in a bay or sea loch for a week or more. During this time the explosive noise of blowing may be almost continuous.

The food of the common porpoise consists chiefly of fish and squids. The fish eaten are small, usually less than 25 cm and rarely more than 35 cm long; they include herring, sprats, mackerel, sand eels, whiting, small cod and hake. A porpoise in pursuit of prey produces echo-ranging clicks as it approaches in order to locate it (see also p. 184). The porpoise can dive to depths of 20 m and remain below the surface for up to 12 minutes.

The male reaches sexual maturity when about 133 cm long and 39 kg in weight; the female is mature at 145 cm when probably 3 – 4 years of age. Mating occurs in the autumn in Europe. During courtship they swim in pairs, close to each other, and approach while moving. The male presents his abdomen and then touches and rubs the side of the female with his tail fluke. The male then passes beneath the female, while they both emit low-frequency sounds. The gestation period is probably of 9 months' duration, young being born in June and July. At birth the young is 60 – 90 cm long, and one of 79·9 cm weighed 8·2 kg. Birth was first observed in captivity in 1914, in a female caught at Dungeness and then taken to the Brighton Aquarium. Typical labour movements were seen first at 03.15 hours and birth took place at 05.50 hours; the cord was broken after the female had swum 6 m and the placenta was delivered 4½ hours later. The young was, however, stillborn; it was a male 65 cm long and 3·17 kg in weight. A few recently born young have been washed ashore on our coasts. One on July 6 in Scotland was 80 cm long and 8·1 kg in weight, others of similar size have been stranded in July and also September. Lactation lasts 6 – 9 months. Adults can reach a length of 1·8 m, and one of 1·47 m weighed 58 kg.

Stenella styx (Euphrosyne Dolphin) closely resembles the common dolphin (1) in size, about 2 m long, but is distinguished by its colouration. The black upper parts are separated from the white under parts by a line passing from the base of the beak though the eye to the underside at the base of the tail. It is usually regarded as oceanic but four have been stranded on our coasts.

1A COMMON DOLPHIN. leaping
1 COMMON DOLPHIN. suckling her young
2 COMMON PORPOISE

· DOLPHINS

The dolphins on this page belong to the order Cetacea (p. 182) and the family Delphinidae (p. 182).

1 Lagenorhynchus acutus (White-sided Dolphin) is the commonest of the dolphins seen around the Shetland Isles. It frequents deeper water, where it may sometimes be seen in summer, jumping high above the surface. The majority reported in the British Isles have been stranded on the Shetland and Orkney Islands; others have occurred as far south as Wexford in Ireland and Yorkshire in England.

It is found in large schools of perhaps 1000 or more animals, and it feeds upon fish, including mackerel, herring and anchovies, as well as crustacea and molluscs such as squids and whelks.

Gestation lasts 10 months and the young are born in June and July. One of 120 cm, stranded on 18 June in the Orkneys, was possibly one of that year's young. Another, caught in Irish waters in July, was 126 cm long, 24 kg in weight, and the teeth were not cut. Adults can reach a length of about 3 m.

2 Lagenorhynchus albirostris (White-beaked Dolphin) appears to migrate northwards in summer into the North Sea and strandings occur then along the east coast and occasionally along the Scottish coast and the Hebrides. In autumn and winter there is a southward migration. Schools are usually of 20 – 30 animals, but sometimes larger. They feed upon fish including herring, whiting and cod. The longest one stranded measured 3 m and one of 2·8 m weighed 305 kg.

3 Tursiops truncatus (Bottle-nosed Dolphin). These dolphins are occasionally stranded on the south and west coasts of England and a few occur along the Irish coasts. They sometimes ascend rivers and some have travelled up the Thames as far as Kew.

They probably reach sexual maturity at 4 years of age but may not bear young until several years later. Aquarium observations have revealed much information about this species, particularly its reproductive behaviour. A male and female will accompany each other for a prolonged perod of courtship during which time the male will stroke and rub the female with his flippers, dorsal fin, and tail flukes, mouth her, clap his jaws (which is also a threat signal) and emit yelps. These activities occur in various sequences but only if the female responds does copulation follow. Most copulatory behaviour takes place at night or early in the morning. The gestation period lasts for 12 months for the last 7 – 8 months of which the female tends to become more isolated except that she maintains association with one or two other dolphins. She becomes less active, slow, and clumsy with the approach of term. The young is born tail first and the female either whirls or swims rapidly ahead so that, when

parturition is complete in about an hour, the short umbilical cord is pulled taut until it breaks. The placenta is not expelled until several hours later. The dorsal fin is folded before and during birth but stiffens and comes into an upright position a few hours later. The tail flukes are also folded. The young can reach the surface unaided to take its first breath 5 – 6 seconds after birth, and shortly after learns to feed by grasping one of the two nipples, which lie on either side of the genital slit. Abdominal contractions of the mother result in the milk being poured into the calf's mouth. Suckling is of short duration as the calf can only remain submerged for less than 1 minute. A young one sucks 2 or 3 times each hour and will do so through the night but as it grows older it suckles less often, particularly at night. It is weaned at 12 – 18 months but may try solid food earlier. In the first few weeks of life it is not allowed to stray more than some 3 m from its mother and is kept close to her for about 4 months.

Aristotle in the 4th century B.C. described the air-borne squeaks and moans of the dolphin — sounds that can be heard today in a dolphinarium. But men have only recently come to understand this animal's ability to use the echo of its own sounds to determine the location and characteristics of an object (echolocation). This species can produce clicks, whistles, and other sounds when finding an object or pursuing prey such as fish or squids. The rate of clicks produced can be adjusted so that the echo arrives after one emitted signal and before the next, in the inter-click interval. Hearing plays a major role and the auditory range of perception is very considerable. The eye is relatively well-developed and vision is used for orientation both in and above the water. These animals have a large brain and show considerable ability for learning; this is observable in captive animals who perform for man. As in other mammals the body temperature is maintained at about 37°C, irrespective of the temperature of the sea water. These bottle-nosed dolphin can reach a length of 3·9 m and one of 3·4 m weighed 394 kg.

4 Grampus griseus (Risso's Dolphin) has been stranded chiefly on the south and west coasts of England and occasionally on the east coast, but only rarely on the coasts of Scotland and Ireland. A newborn one has been found on the Irish coast in May and other young in August and September. It is found throughout the world, either singly or in small schools of up to 12 animals.

The teeth are usually restricted to 3 – 7 on either side of the lower jaw and occasionally some persist in the upper jaw. It feeds on cephalopods, including cuttlefish, and fish. A length of up to 4 m can be reached, and one of 3·4 m weighed 343 kg.

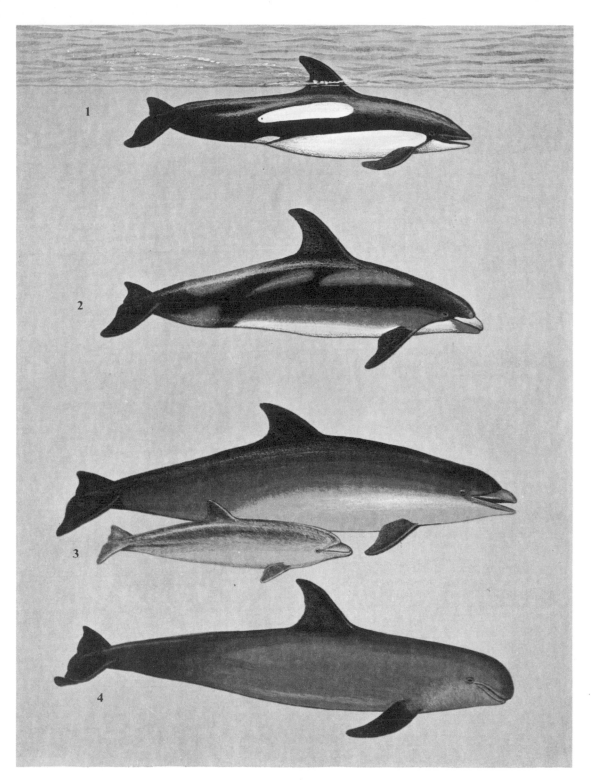

1 WHITE-SIDED DOLPHIN
2 WHITE-BEAKED DOLPHIN
3 BOTTLE-NOSED DOLPHIN, with calf
4 RISSO'S DOLPHIN

185

TOOTHED-WHALES

All the whales on this page belong to the order Cetacea (p. 182); 1–3 belong to the family Delphinidae (p. 182), and *Delphinapterus leucas* and *Monodon monoceros* to the family Monodontidae of which these are the only representatives.

1 Orcinus orca (Killer Whale) is the largest member of the dolphin family and is easily recognized by its colouration. An old male may reach a length of 9 m but females usually attain only 4·5 m. Further sexual dimorphism can be seen in the high dorsal fin, about 2 m in height, of the adult male in contrast with that of the female which is 1 – 1·2 m high.

The distribution of the killer whale is world-wide. There have been some 34 strandings on our shores since 1913, most of which have been along the east coast but a few have occurred on the west coasts and around Ireland. The high dorsal fin is sometimes seen in Shetland waters and once a small school entered Lerwick harbour. It is usually found in small schools consisting of an adult bull and several cows, but lone bulls are sometimes seen. It has a swimming speed of 10 – 13 km per hour.

The most ferocious member of the dolphin family, it is the only cetacean to prey upon other warm-blooded animals, which it takes in preference to fish. It feeds mainly on seals and porpoises but a school will attack a large whale carcass caught by whalers. Penguins and squids are also eaten. The stomach is capacious; one killer whale, 6 m long and 8000 kg in weight, was found to contain 3 porpoises each with a foetus at full term. The mouth is very large and 10 – 14 teeth, each 25 – 50 mm in diameter, are carried on each side of both jaws. The young are born in November and December after a gestation period of about 1 year. One newborn calf found was 2·1 m long and another of 2·7 m was stranded one December in County Clare.

Killer whales have been kept in marine aquaria. One at present kept at Windsor has been trained to perform tasks and can leap out of the water.

2 Pseudorca crassidens (False Killer) is an oceanic species that only rarely comes into coastal waters but when it does so is often stranded in large numbers. One school of some 150 was stranded in Sutherlandshire, and there are other records from various parts of the world. Strandings have occurred in Wales, and along the North Sea coast of Scotland and England as far south as Lincolnshire.

Males can reach a length of 5·6 m and females 5 m; one specimen of 5·3 m weighed 1700 kg. The young are 1·7 – 2 m long at birth.

3 Globicephala melaena (Pilot Whale) occurs in schools usually of 11 – 100 individuals but some of several hundred have occasionally been seen in the vicinity of the Orkney and Shetland Islands — one of 200 and another of 400 – 500 individuals were seen in Lerwick harbour in 1949. They have been stranded, sometimes in large numbers, chiefly along the west coasts and also along the south coast of Ireland. In November 1965 some 63 were stranded at Cloghane, Brandon Bay, in County Kerry and there are a few other similar records. It was hunted till recently in the Orkney and Shetland Islands (and still is in the Faroes) where numbers were driven ashore into suitable places for slaughter and their retreat cut off by a line of boats. The meat and blubber were important in the economy of these islands. This whale was hunted by Stone Age man in northern Europe.

Males are sexually mature when about 4·7 m long and 12 – 13 years of age, and females when some 3·5 m long and 6 years of age. They breed during the winter and spring. The gestation period lasts 15 – 16 months and the young are born over a 6-month period with a peak in August. The calf at birth is 1·5 – 1·8 m long and is suckled for up to 22 months. The eruption of teeth begins when the calf is about 2·13 m and is complete by the time it is some 2·74 m long. The calf begins to take squids when it is about 2·3 m long and probably between 6 and 9 months of age, although it is still suckling. The female does not usually conceive whilst her last offspring is still suckling, and the whole reproductive cycle may therefore occupy as much as 38 months. It has been suggested that the female does not reproduce after 18 years of age; thus the total number of young per female is not very large. One of the largest members of the dolphin family, it can reach a length of 8·6 m but of those stranded on our shores the largest recorded was a male 6·5 m long, while the smallest was 1·77 m. It has been estimated that a female of 4 m weighs about 800 kg and a male of 6 m weighs about 2750 kg. The length of life of pilot whales may be quite long as a male of 40 years and a female of 50 years have been found.

Delphinapterus leucas (White Whale, Beluga) was recorded on nine occasions before 1913 and subsequently a young one was taken near the Forth Bridge in 1932 and another was seen in Loch Long, Dunbartonshire in 1965.

Monodon monoceros (Narwhal) is almost exclusively an Arctic species but has been recorded on our coasts five times since 1648 the last occasion being 1949 when two females were stranded at Rainham, Essex. The bull cannot be mistaken because of the very long, single tooth which projects forward some 2·4 m.

1 KILLER WHALE. Male   1A KILLER WHALE. Female

2 FALSE KILLER
3 PILOT WHALE

TOOTHED-WHALES

The whales on this page belong to the order Cetacea (p. 182); 2–5 are members of the family Ziphiidae, small and medium-sized toothed-whales that have a dorsal fin, well developed beak, and two longitudinal grooves on the throat. The sperm whale (1), the largest of the toothed-whales, is a member of the family Physeteridae, characterized by an enormous head and large snout occupied by a reservoir of sperm oil.

1 **Physeter catodon** (Sperm Whale) is the largest of the toothed whales and occurs regularly in the Atlantic to the west of Ireland in the neighbourhood of Rockall and St. Kilda. It usually appears in the summer and is most abundant during August, although perhaps less so now than in previous centuries. This whale is common in equatorial waters; those found in higher latitudes are almost exclusively adult males. In recent years one of 18 m was stranded on one of the Shetland Islands. Others were stranded in Ireland of which the smallest was 10 m long and the largest 19·8 m — apparently the largest on record. It weighs some 52,400 kg.
This whale shows sexual dimorphism, the female reaching a length of 9 m and the male usually 18 m. The female is mature sexually when 8·5 m long. The species appears to be polygamous, the male mating with more than one female. The mating season is long, lasting from August until March, but the majority mate between October and January. The gestation period lasts about 14·5 months and the young are born between early November and early June, with a peak in February and March. At birth the single calf is 3·7 – 5 m long, and one of 4·04 m weighed 774 kg. The calf may be suckled for as long as 24 – 25 months although considerable variation will occur from one individual to another. The female probably rests for about 9 months before conceiving again.
The large, almost square head is a striking feature of the sperm whale. The skull is more distorted, in relation to the usual symmetric organization, than in any other mammal. There is a single nostril, the left one, from which the whale blows forward at an oblique angle (1A). The region of the right nostril has been transformed and expanded into a reservoir, 'the case', which is filled with a substance called spermaceti. Spermaceti is a liquid oil at body temperature which can be poured from the case after the death of the animal, but sets on cooling to a firm white wax. It has been suggested that the spermaceti is important in the control of buoyancy, and perhaps also in the absorption of nitrogen, particularly during prolonged and deep dives. The sperm whale dives to considerable depths and indeed has been found entangled in submarine cables at depths of 1000 m and 1200 m. It also remains below the surface for as long as 50 minutes, after which it spends about 10 minutes at the surface.
In comparison with the large head the lower jaw is elongate and narrow. There are 16 – 30 large, conical teeth along each side of the lower jaw, that fit into sockets on each side of the palate. It feeds on cephalopods, including squids and cuttlefish, and many beaks from these animals have been found in its digestive tract. It may be the irritation produced by these horny beaks that leads to the formation, in the intestines, of the concretion known as ambergris. A large number of sperm whales caught near Iceland were found to have been eating fish, chiefly *Cyclopterus lumpus* but also *Sebastes* and *Lophius piscatorius*. They probably took this prey at depths of about 500 m. Others had been feeding on sharks.
In the early 19th century many American, and some British, French and German whalers scoured the seas for this whale. As it is a slow-moving species, usually swimming at about 2 m per second or perhaps 6 m per second under stress, it could be chased by ships under sail or oar, and when dead it generally floats so that it could be recovered. Sperm whale oil is used in the dressing of leather, as a lubricant, and in the production of soap and cosmetics. Ambergris, which has been known from antiquity, is used in perfumery as a 'fixative', a substance which has the property of retaining fragrance.

2 **Hyperoodon ampullatus** (Bottle-nosed Whale) is an Atlantic species which migrates to the Arctic in summer. During the autumn and winter it returns south and at this time it may pass through our coastal waters in small schools of 4 – 12 animals. This whale is capable of sudden leaps clear of the water surface.
An old male can reach a length of 9 m; an animal of this size may yield 2000 kg of oil and 100 kg of spermaceti. Females may reach a length of 7 m.

3 **Mesoplodon bidens** (Sowerby's Whale) has been stranded on 21 occasions between 1913 and 1966, on the Irish, Welsh, and English coasts and in the Shetlands. It can reach a length of 4·8 m.

4 **Mesoplodon mirus** (True's Beaked Whale), one of the rarest of the beaked whales, has been recorded four times in Ireland, the last in October 1967 when one of about 4 m was washed ashore in County Kerry.

5 **Ziphius cavirostris** (Cuvier's Whale) is an Atlantic species that has been recorded on more than 30 occasions since 1913 on the coasts of Ireland, Scotland, and the west of England. It has been found to feed on cephalopods. It may reach a length of 8·5 m.

2A   1A

1 SPERM WHALE   1A SPERM WHALE, blowing
2 BOTTLE-NOSED WHALE   2A BOTTLE-NOSED WHALE, blowing
3 SOWERBY'S WHALE   4 TRUE'S BEAKED WHALE
5 CUVIER'S WHALE

WHALEBONE WHALES

All belong to the order Cetacea (p. 182) and the sub-order Mysticeti, distinguished by a series of plates, baleen or whalebone, which take the place of teeth in the upper jaw. The frayed inner edges of these plates act as filters for the small planktonic organisms on which these whales feed. With the exception of *Balaena glacialis* all are members of the family Balaenopteridae which includes the largest animal that has ever lived (1); they have a series of parallel grooves below the throat. The family Balaenidae, to which *Balaena glacialis* belongs, have no such grooves and a much larger mouth.

1 **Balaenoptera musculus** (Blue Whale, Sibbald's Rorqual) is the largest animal that has ever existed. It can reach a length of 30 m, and a weight of 120,000 kg — equivalent to some 24 elephants. It has a world-wide distribution and migrates between high and low latitudes. During the northward feeding migration it often reaches the latitude of the Shetlands in July or August. It is usually seen off-shore singly or in pairs, rarely in schools. It can swim at about 7 m per second for 2 hours, but at 10 m per second for only 10 minutes. Four were stranded on British coasts between 1913 and 1966.

In contrast with its enormous size it feeds upon tiny planktonic crustacea, the krill, chiefly the shrimp-like euphausid *Boreophausia inermis* in the northern hemisphere. The krill is collected on the fibrous fringe of the horny baleen plates. These hang from the margin of the upper jaw, 300 - 400 plates on each side. The longest of them is about 1 m. To feed, the whale opens its mouth, takes in a large volume of water, closes its mouth, and with its tongue forces the water out through the fringe so that the plankton is filtered off and retained. The stomach of one specimen, 26 m long, was estimated to contain 2000 kg of krill. The throat has some 70 - 120 grooves which are capable of expansion.

The male is sexually mature when about 22·5 m in length, and the female when about 5 years of age and 23·8 m in length. Pairing probably occurs in June and July. The gestation period lasts 10 - 11 months; during the last 2 months the foetus gains about 2000 kg of its birth weight of some 2500 kg. The calf, born in the warmer waters of lower latitudes, is about 7 m long at birth; one measured 7·46 m and weighed 2794 kg. Lactation lasts 6 - 7 months. Many females probably conceive every 2 years, others every 3 years. Pregnant females 30 years of age have been found and probably blue whales can survive longer than this.

Its long gestation period, and consequent low rate of reproduction together with the large numbers that have been taken by man in the past century have greatly reduced its chance of survival.

2 **Balaenoptera physalus** (Common Rorqual, Common Fin Whale) has a world-wide distribution and is next in size to the blue whale, reaching a length of 24 m and a weight of some 70,000 kg. These whales have been stranded along the coast between the Shetlands and Cornwall and all around Ireland, usually either during late winter and spring on their journey northwards or in autumn when going southwards. The male is mature sexually when about 19·2 m long and the female when 19·8 m. Pairing occurs in June and July and gestation lasts 11 months. The calf at birth is about 6 m long and weighs some 1800 kg. Like the blue whale, it feeds mainly on small planktonic crustacea about 9 mm long; but it will also eat herrings. The baleen plates show asymmetry in colouration; all are dark slate with longitudinal yellowish streaks with the exception of the front one-third on the right which are white or yellowish-white.

3 **Balaenoptera acutorostrata** (Lesser Rorqual, Piked Whale) has been stranded all around the British Isles except the southern North Sea and the English Channel. It occurred frequently in the inshore waters of the Shetlands until about 1945. The smallest of the fin whales, it can reach a length of 9 m and is sexually mature when about 7·4 m long. The gestation period lasts 10 - 11 months and the single calf, about 3 m long, is usually born between November and February. Lactation lasts 4 - 5 months. It is said to eat small fish as well as planktonic organisms.

Balaenoptera borealis (Sei Whale or Rudolphi's Rorqual) has a world-wide distribution, and although rarely stranded around the British Isles has often been seen off-shore from the north-west Highlands of Scotland. Similar but less slender than the fin whale, it can reach a length of 18 m, but is usually 14 m long and 49,000 kg in weight. The male is mature sexually when about 13·4 m long and the female when 14·4 m. Pairing occurs between June and August and gestation lasts 11 - 12 months; lactation lasts 4 - 5 months. Very small planktonic animals, particularly crustacea, are their chief food.

Balaena glacialis (North Atlantic, Biscayan Right Whale) is the only one of the right whales recorded in British waters. It used to be caught commercially off the Hebrides. 'Right' whales were so called by whalers who could capture them easily because they moved slowly and floated when dead. This whale can reach a length of 18 m. Some 230 black baleen plates are present, the largest being 1·8 - 2·7 m long.

Megaptera novaeangliae (Humpback) has not been recorded since 1913 but whalers used to take it in British waters. It may reach a length of 15 m and has long flippers, 3 - 4 m. Males of about 11·8 m and females of 12·4 m are sexually mature.

1 BLUE WHALE
2 COMMON RORQUAL. with young   2A COMMON RORQUAL. spouting
3 LESSER RORQUAL

THE VERTEBRATES: A CLASSIFICATION

As there are approximately one million known living species of animals, as well as innumerable known fossil forms, it is clearly essential to try to arrange such an assemblage in some sort of order. The method in use today is based upon that of the Swedish biologist, Linnaeus, who in the eighteenth century grouped together animals with similar morphological characteristics. Subsequently this classification has been greatly modified as more information has become available from living animals as well as from fossil forms whose significance was not appreciated at that time.

Linnaeus used two names for each of the then known animals. The first term is that of the genus and the second is the specific one, giving together the name of the species. This is a group of individuals able to breed amongst themselves but not with other species. Within a species there are often various subspecies (races) and these may become species by reproductive isolation, for example island communities. The subspecies is denoted by three terms. A number of closely related species are grouped into a genus and these in turn into larger groups. In ascending order we have therefore Subspecies, Species, Genus, Family, Order, Class, Phylum and Kingdom. Some of these groups may be further subdivided, and in this book we are concerned with the Subphylum Vertebrata.

The classification scheme used here follows that given by A. S. Romer in his valuable book *Vertebrate Paleontology* (3rd edition, 1966; The University of Chicago Press). In the following synopsis all the Classes of the Subphylum Vertebrata have been included. Those which contain only fossil forms or animals normally found in other regions are shown in parentheses. All the genera found in the British Isles have been included, as well as introduced ones. Names of the genera and species of fishes follow those given by A. Wheeler in *The Fishes of the British Isles and North-West Europe* (1969; Macmillan, London). Those for the amphibia and reptiles follow M. Smith in *The British Amphibians and Reptiles* (4th edition, 1969; Collins, London), while those for the mammals can be found in a small handbook *The Identification of British Mammals* by G. B. Corbet (1964; The Trustees of the British Museum, Natural History).

Subphylum **VERTEBRATA**

Vertebrates. Animals which develop vertebrae and whose nervous system is differentiated anteriorly into an elaborate brain, housed in a cranium.

Class **AGNATHA**

Earliest vertebrates characterized by the absence of jaws. (Only one order is mentioned below, the others being fossil orders.)

Order **Cyclostomata** Lampreys and hagfish. Eel-shaped but lack paired fins; skeleton of cartilage; pouch-like gills.

| *Petromyzon* | *Lampetra* | *Myxine* |

[*Class* **PLACODERMI**]

The earliest vertebrates with jaws. Flourished in the Devonian and are known only from fossils.

Class **CHONDRICHTHYES**

Fish with jaws; skeleton of cartilage; paired pectoral and pelvic fins.

Subclass Elasmobranchii Sharks and rays. Five to seven pairs of gills; numerous teeth. (Not all orders mentioned here.)

Order **Selachii** Sharks. Spiracle and gill-slits at side of head; pectoral fins free from head. Most are good swimmers and majority are predators.

Hexanchus	*Chlamydoselachus*	*Isurus*
Lamna	*Cetorhinus*	*Alopias*
Galeus	*Scyliorhinus*	*Mustelus*
Galeorhinus	*Prionace*	*Sphyrna*
Oxynotus	*Squalus*	*Dalatias*
Somniosus	*Echinorhinus*	*Squatina*

Order　**Batoidea**　Skates and rays. Bottom-living and often very depressed dorso-ventrally; anal fin absent; laterally expanded pectoral fins extend forward to join side of head; large spiracle dorsally placed and gills ventrally; eyes dorsal.

Torpedo	*Raja*	*Dasyatis*
Trygon	*Mobula*	*Myliobatis*

Subclass　HOLOCEPHALI　Chimaeras or ratfish. Gill-clefts open into a single chamber with single external opening; small mouth with lips; upper jaw fused to skull; tail fin often reduced to whiplash. Fairly active swimmers. Contains only one order—Chimaeriformes.

Chimaera

Class　OSTEICHTHYES

Skeleton of bone; gill-slits covered by an operculum; paired pectoral and pelvic fins. (Not all the orders are mentioned below.)

Subclass　ACTINOPTERYGII　Ray-finned fish; nostrils generally without internal opening and situated well up on front of head.

Infraclass　**Chondrostei**　Most are known from fossils. Early forms and a few surviving ones have thick scales; uptilted tail; prominent snout.

Order　**Acipenseriformes**　Sturgeons. Body naked except for 5 longitudinal rows of bony scutes; mouth rather small with a row of barbels in front of it.

Acipenser

[*Infraclass*　**Holostei**]　Known mainly from fossils. Dominant fish in the middle Mesozoic; the few living forms include the garpike and bowfin both found in fresh-waters of North America.

Infraclass　**Teleostei**　Modern fish, dominant group since the Cretaceous; tail superficially symmetric but internally uptilted. (Some orders have been omitted.)

Super-order　**Elopomorpha**　A primitive group in which there is a leptocephalus larva.

Order　**Anguilliformes**　Eels. Morphologically separated from other teleost orders. Body long and slender; pelvic fins absent in modern forms; scales absent or rudimentary. Mainly marine but some, like *Anguilla*, spend adult life in freshwater.

Anguilla	*Muraena*	*Conger*

Super-order　**Osteoglossomorpha**│　Freshwater fish confined to the tropical regions of South America, Africa, South-east Asia, New Guinea and Australia, except for one North American species. Includes *Mormyrus* the elephant-snouted fish of Africa.

Super-order　**Clupeomorpha**　A primitive group but successful in modern times. Silvery fish with easily lost scales; lateral line canals on head extend over operculum; usually no lateral line pores on body.

Order　**Clupeiformes**　Herring and relatives. Only order of the above super-order.

Alosa	*Clupea*	*Engraulis*
Sardina	*Sprattus*	

Super-order　**Protacanthopterygii**　Majority are slender predatory fish of salt and fresh water; upper jaw slightly protrusible in a few species. (Only one of the four orders is mentioned below.)

Order　**Salmoniformes**　Principal group of the super-order. The maxilla is excluded by the premaxilla from the gape.

Suborder　**Salmonoidei**　Salmon and relatives. All have adipose fin.

Coregonus	*Salmo*	*Salvelinus*
Thymallus	*Osmerus*	

Suborder　**Argentinoidei**　Argentines. Elongated silvery fish with large eyes and small mouth.

Argentina

Suborder　**Esocoidei**　Pike. Small group of northern freshwater fish.

Esox

Suborder　**Stomiatoidei**　Oceanic species with organs which emit light.

Maurolicus

Super-order **Ostariophysi** Predominantly freshwater fish of diverse form and habit; includes predators, herbivores, detritus and microphagous feeders. May have a protrusible upper jaw and many have well-developed circumoral barbels. Pelvic fins abdominal; fin spines often present. Many groups with adipose fin.

Order **Cypriniformes** Carp and relatives. Body usually has scales, only rarely naked; no vomerine teeth.

Abramis	*Alburnus*	*Barbus*
Blicca	*Carassius*	*Cyprinus*
Gobio	*Leuciscus*	*Phoxinus*
Rhodeus	*Rutilus*	*Scardinius*
Tinca	*Cobitis*	*Noemacheilus*

[*Order* **Siluriformes**] Catfish. The majority are found in South America and Africa; *Silurus* was introduced to British Isles from Europe. Body naked or covered with bony plates; some species have spinules or prickles. Adipose fin usually present; several pairs of barbels, which are their most striking feature.

Silurus

Super-order **Paracanthopterygii** Mainly marine, some members live in deep water. Stout, soft-bodied fish. Number of species viviparous. All are carnivorous. (Not all orders included.)

Order **Gobiesociformes** Clingfish or suckers. Small saltwater fish. Body is flattened; no fin spines; pelvic fins have moved anteriorly and form a complex suction disc.

Apletodon	*Diplecogaster*	*Lepadogaster*

Order **Lophiiformes** Anglers. First dorsal fin spine carries a tassel to lure prey.

Lophius

Order **Gadiformes** Cod and relatives. Typically long-bodied, tapering posteriorly; dorsal and anal fins are elongated or may be divided into 2 or 3 parts; little development of fin spines.

Brosme	*Ciliata*	*Gadus*
Gaidropsarus	*Lota*	*Melanogrammus*
Merlangius	*Merluccius*	*Micromesistius*
Molva	*Phycis*	*Pollachius*
Raniceps	*Rhinonemus*	*Trisopterus*
Zoarces		

Super-order **Atherinomorpha** Small, surface-feeding fish, mainly fresh or brackish water, some marine species. Pronounced sexual dimorphism often seen in freshwater species. Many viviparous; oviparous species produce large, demersal eggs with adhesive filaments.

Order **Atheriniformes** Flying fish and skippers. The only order in the super-order.

Exocoetus	*Belone*	*Scomberesox*
Atherina		

Super-order **Acanthopterygii** Spiny-finned teleosts. This super-order includes the great majority of modern marine, benthic and littoral species, and a few freshwater forms. Various foods utilized, with corresponding diversity in feeding mechanisms; upper jaw protrusible in many species. Development of stiff fin spines; operculum often armed. Majority oviparous. (Not all orders are mentioned below.)

Order **Zeiformes** John Dory and relatives. A small group of deep-bodied, laterally compressed fish with large eyes. First dorsal fin with spiny rays.

Zeus

Order **Lampridiformes** Moonfish and relatives. Live in upper waters of open sea.

Lampris

Order **Gasterosteiformes** Sticklebacks, sea horses and pipe-fishes. Small fish usually with elongate body encased in bony armour.

Gasterosteus	*Pungitius*	*Spinachia*
Entelurus	*Hippocampus*	*Nerophis*
Syngnathus		

Order **Scorpaeniformes** Gurnards, bullheads and relatives. Mail-cheeked fish, so-named from the specialized cranial structures; large pectoral fins.

Aspitrigla	Eutrigla	Sebastes
Trigla	Trigloporus	Cottus
Myoxocephalus	Taurulus	Agonus
Cyclopterus	Liparis	

Order **Perciformes** In early Tertiary this group showed rapid radiation. Great majority of spiny-finned fish regarded currently as belonging to this order.

Suborder **Percoidei** Perch and relatives.

Dicentrarchus	Epinephelus	Morone
Polyprion	Serranus	Micropterus
Gymnocephalus	Perca	Stizostedion
Remora	Naucrates	Seriola
Trachurus	Brama	Boops
Dentex	Pagellus	Pagrus
Sparus	Spondyliosoma	Mullus
Cepola		

Suborder **Mugiloidei** Mullets and relatives. Pelvic fins are somwhat posterior.

Crenimugil	Liza	Mugil

Suborder **Labroidei** Wrasse and relatives. Moderately deep-bodied fish, often brightly coloured; small mouth but powerful dentition.

Centrolabrus	Coris	Crenilabrus
Ctenolabrus	Labrus	

Suborder **Trachinoidei** Weevers. Usually small fish with long body tapering posteriorly; long dorsal and anal fins; eyes dorsally placed.

Trachinus

Suborder **Blennioidei** Blennies and relatives. Small fish that dwell along rocky shores. Elongated tapering body; long dorsal and anal fins; small pelvic fins have moved forward, pectoral fins large.

Blennius	Coryphoblennius	Chirolophis
Lumpenus	Pholis	

Suborder **Ammodytoidei** Sand eels. Small slender fish that bury themselves in sand; no spines in dorsal fin.

Ammodytes	Gymnammodytes	Hyperoplus

Suborder **Callionymoidei** Dragonets. Head flattened; scales absent.

Callionymus

Suborder **Gobioidei** Gobies. Small fish, some common near shore. Typically the pelvic fins are fused by the fin membrane to form a suction organ.

Aphia	Buenia	Chaparrudo
Crystallogobius	Gobius	Lebetus
Lesueurigobius	Pomatoschistus	

Suborder **Scombroidei** Mackerel, tunny and relatives. Streamlined fusiform body; forked tail; row of dorsal and anal finlets. Fast-swimming surface dwellers of the oceans.

Auxis	Euthynnus	Katsuwonus
Sarda	Scomber	Thunnus

Order **Pleuronectiformes** Plaice, turbot, halibut, sole and relatives. Flatfish which have settled to the bottom on one side, producing asymmetric development of head, eyes and jaws.

Arnoglossus	Buglossidium	Glyptocephalus
Hippoglossoides	Hippoglossus	Lepidorhombus
Limanda	Microchirus	Microstomus
Pegusa	Phrynorhombus	Platichthys
Pleuronectes	Scophthalmus	Solea
Zeugopterus		

Order **Tetraodontiformes** Varied group of fish. All have tiny mouth armed with heavy teeth; powerful jaws; upper jaw fused to brain-case; short deep body; spiny dorsal fin (often lost) and also pelvic fins.

Suborder **Balistoidei** Trigger and file fishes. Deep-bodied, flat-sided fish in which spiny dorsal fin generally reduced to one large upward-projecting spine.
Balistes

Suborder **Tetraodontoidei** Puffer fish and sunfish. Deep-bodied fish.
Lagocephalus   *Mola*

[*Subclass* SARCOPTERYGII] Fleshy-finned fish of great evolutionary interest as it was from this stock that the land vertebrates were derived. Today represented by the coelacanth, *Latimeria*, and the lungfishes. The coelacanth, although known from fossils, was considered extinct until 1938 when a live one was caught in deep water off Africa. During their evolution coelacanths moved from fresh water to salt and finally to deep-sea. The lungfishes are found in fresh water in tropical areas of Australia, Africa and South America. They obtain nearly all their oxygen from the air through their paired lungs.

Class **AMPHIBIA**

The most primitive land vertebrates, and the first group of vertebrates to emerge from water. They have four limbs although there is reduction in some cases. Most amphibia breed in water; the larvae typically have gills for respiration in water and the adults lungs for air-breathing. Skin is typically without scales.

Subclass LISSAMPHIBIA Includes all modern orders of amphibia. (Not all orders given here.)

Order **Anura** Frogs and toads. Long hind limbs for jumping; tail present in larval stage when aquatic but absorbed about the time it leaves the water.
Alytes   *Bufo*   *Hyla*
Rana

Order **Urodela** Newts and salamanders. Retain post-anal tail; limbs similar in length.
Triturus

Class **REPTILIA**

More fully terrestrial, apart from some species such as sea-turtles, than amphibia. Skin usually covered with horny scales; lungs for air-breathing. Fertilization is internal; either viviparous or oviparous; shelled eggs, containing large yolk, are laid on land; no larval stage. (Not all orders have been included here.)

Subclass ANAPSIDA Skull is without temporal openings.

Order **Chelonia** Turtles. The turtle shell, defensive armour, is characteristic.
Caretta   *Lepidochelys*   *Dermochelys*

Subclass LEPIDOSAURIA 'Scaly' reptiles. Skull has two temporal openings, above and below post-orbital and squamosal bones.

Order **Squamata** Lizards and snakes, the most successful of modern reptiles.

Suborder **Lacertilia** Lizards. Generally with four limbs, but some are snake-like. Typically, visible eardrum and movable eyelid.
Lacerta   *Anguis*

Suborder **Ophidia** Snakes. Typically, all traces of limbs lost. Skull modified to provide greater motility of its parts for swallowing large prey.
Coronella   *Natrix*   *Vipera*

[*Subclass* ARCHOSAURIA] 'Ruling reptiles', chiefly fossil. The crocodiles and alligators survive today.

[*Subclass* SYNAPSIDA] Fossil forms leading to mammals.

[*Class* **AVES**]
See *Oxford Book of Birds*

Class **MAMMALIA**

Most successful of the land vertebrates. Brain much enlarged; teeth only along margin of jaws; hair on body, constant body temperature; young suckled after birth.

[*Subclass* PROTOTHERIA] Egg-laying mammals, the monotremes, found in Australia; the duck-billed platypus and the spiny anteater survive today.

Subclass THERIA Bear their young alive.

Infraclass **Metatheria** Pouched mammals, found mostly in Australia, a few in America. Young, born at an immature stage, are carried and suckled in pouch after birth.

Order **Marsupialia** Marsupials. Constitute the greater part of the fauna in the Australian region. One species, *Macropus rufogriseus*, has been introduced and is now breeding ferally in Sussex, Derbyshire and Staffordshire.
>> *Macropus*

Infraclass **Eutheria** Higher mammals. Have efficient placenta and the young are born at a more advanced stage than in Metatheria. (Not all orders included below.)

Order **Insectivora** Insectivores: hedgehog, moles, shrews and their relatives. Generally small in size; snout usually elongated and tapering.

Erinaceus	*Talpa*	*Crocidura*
Neomys	*Sorex*	

Order **Chiroptera** Bats. Capable of true and sustained flight. Fore-limbs have four elongated digits; a membrane is stretched between these digits, the hind limb, and the tail; the thumb remains free and is clawed.

Rhinolophus	*Barbastella*	*Eptesicus*
Myotis	*Nyctalus*	*Pipistrellus*
Plecotus	*Vespertilio*	

Order **Carnivora** Carnivores or flesh-eaters. Teeth adapted for biting and tearing flesh.

Suborder **Fissipedia** Land carnivores: otter, badger, fox, cat and relatives. Musk gland at base of tail.

Lutra	*Martes*	*Meles*
Mustela	*Vulpes*	*Felis*

Suborder **Pinnipedia** Marine carnivores: walrus and seals. Streamlined, torpedo-shaped body; limbs modified to form flippers for swimming, the digits being broadly webbed.

Odobenus	*Cystophora*	*Erignathus*
Halichoerus	*Phoca*	

Order **Perissodactyla** Odd-toed ungulates: tapirs, rhinoceroses and horses. In some the limbs have become long and slim, the distal segments having lengthened in comparison with the proximal ones; digits reduced to three or even one. Herbivores with teeth highly adapted for grinding.
>> *Equus*

Order **Artiodactyla** Even-toed ungulates: a variety of hoofed mammals, including pigs, hippopotami, cows, sheep, goats, deer, antelopes and relatives. Third and fourth digits well-developed.

Capreolus	*Cervus*	*Dama*
Hydropotes	*Muntiacus*	*Rangifer*
Bos	*Capra*	*Ovis*

Order **Cetacea** Whales, dolphins and porpoises. Have become wholly aquatic and are helpless if stranded; only their need to breathe air indicates their earlier terrestrial existence. No external hind limbs; fore-limbs have developed into flippers; tail has horizontal flukes.

Suborder **Odontoceti** Toothed whales; includes the majority of living whales. The teeth, although occasionally concealed, are present in the gums. Single blowhole.

Hyperoodon	*Mesoplodon*	*Ziphius*
Delphinus	*Globicephala*	*Grampus*
Lagenorhynchus	*Orcinus*	*Pseudorca*
Stenella	*Tursiops*	*Delphinapterus*
Monodon	*Phocoena*	*Physeter*

Suborder **Mysticeti** Whalebone or baleen whales. Small number of types, all enormous, one being the largest of all vertebrates. Paried blowhole; baleen plates, suspended in mouth, filter food from sea water.

Balaenoptera	*Megaptera*	*Balaena*

Order **Rodentia** Squirrels, rats, mice and relatives. Gnawing animals of small or moderate size. Two upper and two lower incisors, covered with enamel on front only.

Suborder **Sciuromorpha**   Squirrel and relatives. Arboreal in habit.
Sciurus

[*Suborder* **Caviomorpha**]   South American group; includes guinea pig and coypu (introduced).
Myocastor

Suborder **Myomorpha**   Rats, mice, dormice, and relatives. Premolars have been lost leaving three molars, upper and lower, to grind food.

Arvicola	*Clethrionomys*	*Microtus*
Ondatra	*Apodemus*	*Micromys*
Mus	*Rattus*	*Glis*
Muscardinus		

Order **Lagomorpha**   Hares and rabbits. Moderate-sized gnawing animals; four incisor teeth plus two accessory teeth behind the upper incisors; tail short or absent.
Lepus   *Oryctolagus*

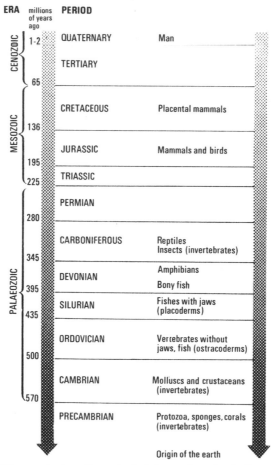

Diagram to show the geological time scale and to indicate the period in which the first fossils of the main groups of animals appeared. Simplified from *An introduction to the study of man* by J.Z.Young, 1971. Clarendon Press, Oxford

THE VERTEBRATES OF THE BRITISH ISLES

Vertebrates of the surrounding seas

The British Isles, today surrounded by sea, once formed part of Europe, the last land connections being severed about 7,000 years ago. Ireland may have remained joined to Scotland later than this. The sea floor around the British Isles forms part of the continental shelf, which extends from the coastline to the shelf edge at a depth of about 200 m (Fig. 1). Most of the shelf has very gentle relief, but there are steep, rocky slopes close inshore and sandbanks and deep holes further out. The bottom is generally formed of mud, sand or gravel and much of this sediment is subject to frequent movement by tidal currents and waves; hard rock occurs over restricted areas. Beyond the shelf edge the continental slope, dissected by submarine canyons, plunges steeply to great depths 4,000 m and more, off the south-western approaches to the British Isles. Much of the deep ocean floor is formed by flat abyssal plains, but volcanic hills also occur. The Faroe Islands, formed of volcanic lavas, are peaks of the Wyville Thomson Ridge which separates the deep waters of the Atlantic and Arctic Oceans (Fig. 2). The deep water faunas on each side of the ridge are different, that to the south-west being richer in variety than that of the colder north-east side.

The marine environment when seen from the edge of the sea appears superficially to be simple, for the waves break with some regularity along the shore and the water may be grey, greenish-blue or blue. But the sea is not really homogeneous at all and many factors combine to give it varying physical features. These include water movements, pressure, temperature, light and salinity. One of the most easily measured of these is depth and it is often used as a basis for classification (Fig. 1). The most densely populated zone of the sea is that between the surface and the depth to which light can penetrate. This is the only region in which plants can live and manufacture their foodstuff. They then form a rich food supply for many animals, among them crustaceans, tunicates, molluscs, and the larvae of many groups including fish. In their turn these animals are eaten by larger predators including herrings, skippers, basking sharks, flying fish and the enormous baleen whales. Mackerel sharks, tunny, and toothed whales that prey on other fish are also found in this zone.

Fig. 1

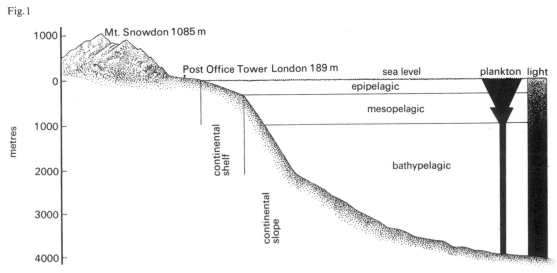

Diagram of a section through land and sea. The right-hand column indicates the depth to which light can penetrate; the other column shows the density of plankton at different levels.

Fig. 2   The British Isles and their relationship to the continental
           shelf edge, the Faroe Islands, the Wyville Thomson Ridge,
           and the Atlantic and Arctic Oceans.

Animals of the upper layers of the oceans tend to be transparent or blue, while those living a little deeper, particularly the fish, are often silvery or greyish in colour. The upper surface of many fish is dark grey or blue and the lower one dull white, while the sides and belly are silvery; others have reflecting areas, the back again being dark while the sides and belly have reflectors. This countershading provides protection as the fish seems to disappear into the background. In the depths where no light penetrates animals are often unusual in shape and many have their own sources of light. The continental slope is inhabited by deep water, bottom-living animals and amongst the fish found here are the black-mouthed dogfish, darkie charlie, the rabbit fish and many others. Some of the fish of this region are not well known as special equipment is needed to capture them.

The extensive coastline of the British Isles encircles many hundreds of islands that vary in size from a few square kilometres to the two large islands of England, Wales and Scotland and of Ireland. The coast varies from steep cliffs to expanses of flat sands with many intermediate types. They are rich in animals which have special modifications in behaviour or in body form that allows them to exploit this habitat and survive the actions of the tides and waves. On rocky coasts the fish often have flattened heads and bodies, like the sea scorpions, while others have developed suckers so that they can retain their position on a rock, like the lumpsuckers and sea snails. Sandy coasts have their own distinctive fauna including the sand eels which can bury themselves at low water, the weevers, the sand gobies and flatfish. The grey and common seal also come ashore, particularly during the breeding season.

Another habitat is that of estuaries where fresh and salt water mixes. Amongst the fish found in these brackish waters are grey mullets, flounders, and bass as well as salmon and eels during their migratory journeys.

200

Freshwater vertebrates

The large islands consisting of England, Wales and Scotland and of Ireland can be divided roughly into the highlands, north-west of a line drawn between the mouths of the rivers Tees and Exe, and the lowlands to the south-east. The scenery and form of the land is determined by its underlying rocks, or geological structure, and by climate. The highlands contain the highest peaks and consist largely of rocks geologically older than those of the lowlands, where there are low hills and broad valleys. A lot of rain falls over the highlands and there are many swift-running streams and rivers while the lowlands have less rain and the rivers are generally meandering and slow-flowing.

During the ice ages, or glaciations, of the last million years ice sheets at times covered the whole of the British Isles with the exception of southern England, along the Thames valley to Gloucester, and the extreme south-west of Ireland. Each ice sheet descended from highland areas and later returned towards them, so that highlands were ice-covered longer than lowlands. At such times the ice-free areas must be envisaged as showing tundra conditions, and the retreating ice must have left behind a bare, devastated landscape capable at first of supporting only a limited amount and variety of life. Interglacial times, on the other hand, were sometimes warmer than the present climate. Each glaciation forced many animals to leave and some species never returned. Many mammals lived in this country during the glacial and inter-glacial periods, which have since left or become extinct. These include the mammoth, woolly rhinocerus, hyaena and bears during the glaciations, and straight-tusked elephant and hippopotamus during the interglacial periods, to name only a few. Between the last ice retreat and the time when the connection with Europe was broken, a period of perhaps some 3,000 years, animals were able to pass freely to and from the continent. Today the preponderance of species of land vertebrates in the south-east reflects these migrations. The rest of Europe contains many more vertebrate species than are found in England, Wales and Scotland, while the situation in Ireland is considerably worse, as few vertebrate animals had reached there before the land connections were severed.

The advancing ice sheets spared the rivers in the southern part of England. It is possible that chars and whitefish were able to populate them although they probably only entered rivers to breed at this time as do the salmon and sea trout today. The char and whitefish would then have been able to colonize the rivers further north as the ice retreated, and later the lakes where they became isolated and are found today. Other species of fish entered rivers in the south as the climate mellowed, arriving either by sea or through river basins, for example the Thames which was then a tributary of the Rhine. These events are reflected by the presence in rivers of south-east England of species of fish that are absent further north. The habitats of a river system vary in many ways and determine the species that survive in them. The amount of dissolved oxygen in the water is critical for many fish, for example the trout and salmon parr which can survive only in swift-flowing rivers and streams. Where the flow is slightly less rapid, bullheads, stoneloach and minnows may be found. In lower reaches, where the flow is only moderately fast, chub, dace, gudgeon and bleak predominate. Slow-flowing waters generally contain tench, bream, carp, roach and rudd together with predators like the pike, perch and eel.

Other freshwater habitats include ponds and lakes and these may vary in area, depth and form with resultant differences in the animals found in them. Some lakes in mountainous areas have steep sides, very deep cool water, and are without shallow regions. The fish in such a lake include trout, whitefish and char. Less deep lakes which also have some shallow water generally contain pike, perch, trout and sticklebacks. Shallow lakes, associated with hilly regions, usually contain more vegetation and the characteristic inhabitants are bream, carp and tench. Of our few native amphibians some are still restricted to the south. They usually live in ponds or the shallow regions of lakes during the breeding season. Much of the rest of the year is spent in surrounding areas. Our water-living mammals include the otter, water-vole and water shrew, together with the coypu which has recently been introduced.

Land vertebrates

Of the completely terrestrial vertebrates the reptiles form a very small group and some are still restricted in their distribution. The sand lizard is found in England and Wales, the slow worm throughout England, Wales and Scotland, whilst the viviparous lizard is also present in Ireland, where it is the only reptile. Of the snakes the smooth snake is found in the south of England, the grass snake in England and Wales, while the adder has penetrated much further as it is also present in Scotland. The land mammals of the British Isles have arrived by various means, the majority probably as immigrants when land connections with Europe existed. Others have been introduced either accidentally or deliberately by man, for example the rabbit in the twelfth century and more recently the grey squirrel and the coypu. Some have arrived on transportation devised by man, like the brown rat. Amongst the mammals only the bats are capable of sustained flight and some of them have crossed the barrier of the English Channel — indeed Nathusius's pipistrelle bat, first reported in the south of England in 1969, is a very recent arrival from Europe.

In the temperate climate of the British Isles the mammals have invaded many of the habitats available. This is due to several factors including their ability to maintain their internal environment at a fairly constant temperature and their powers of locomotion. Other factors, however, may tend to prevent their spread: these include the availability of food; the competition with other animals for living space; predators; water barriers; and mountainous areas where sharp contours make for radical changes and where some animals are unable to breed. Although mammals adapt to many types of habitat, there is usually one in which a species is most successful and with which it is associated. The brown hare, for example, usually lives on grassland, the otter in marshes or the bogs of river systems, the short-tailed vole in scrub or grassland, and the dormouse in deciduous woodland. Animals which can adapt to different habitats and diets and reproduce under varied conditions are likely to increase their population and to spread, particularly if the rate of reproduction is high. Of the mammals the rodents are probably the most successful. A measure of this was the occupation of a refrigerated store by house mice which lived in total darkness at a temperature of $-10°C$ and fed only on the meat kept there. Their rate of reproduction was slightly better than the average for house mice.

The complete history of the mammals of these islands is not known, as the fossil as well as the more recent history is needed for each species. However, it is possible to trace the more spectacular events in the history of some animals. For instance the sea has formed an effective barrier to the mole, the common shrew and the water vole as these animals did not succeed in reaching Ireland before the land connections were broken. The sea separating the small islands from the larger land masses, although shallow in most places, has proved a formidable barrier and where mammals have invaded an island it has usually been with the help of man. Where island populations exist, then subspecies, or even species, may develop in isolation. An example of this is the house mouse which was taken to St. Kilda by the first human settlers, within historic times, and which has since developed into a subspecies *Mus mus muralis*.

The distribution of animals

If the distribution and population of each species of animal were known this information could be built into a series of contour maps. Each peak would indicate an area where a large population of one species was found at the time of recording. The distribution of animals in these islands is not well known, the species present in one region often being better known for some places than for others, usually due to the presence of a biological station or an active natural history society. Today, however, with the advent of a scheme to collect records of animals present in 10 km squares of the National Grid, it is possible to compile distribution maps. This is organised by the Biological Records Centre and maps have recently been published for the land mammals[1] and for some of the freshwater fish[2]. Records are being collected for the shore fish and also for the invertebrate species. These detailed records will be extremely valuable, not only for revealing the present day distribution of animals, but also for comparison with similar records taken in the future. They will help us to understand the multitudinous effects both of the environment and of other animals and man on our heritage of wildlife. It is hard to predict the future for the animals of these islands, but perhaps some salutary lessons may be learned from changes in the population of some of the mammals in recent times. One example is the effect of myxomatosis, a virus disease, on the rabbit during the 1950s. Within 20 years there are signs of recovery because resistance to the disease has developed. Another example is the development of some strains of rat that are resistant to the poison Warfarin. Some introduced species have succeeded in breeding and spreading, probably at the expense of other species, the grey squirrel and the coypu being examples.

In other ways, perhaps less obvious, man is disturbing the flora and fauna of these islands. Buildings, towns, motorways and other concrete structures may provide niches for birds, bats and rats but the vegetation, food, cover and habitats of others is being destroyed or reduced. In rich agricultural areas hedgerows are being destroyed to permit the more efficient use of expensive farm machinery and labour. This results in the loss of habitats for many small animals and plants. As time for leisure increases, the people from towns will spend more time in visiting and enjoying the country. Care is needed to ensure that this great amenity is not destroyed, even gradually. The New Forest remains at present but already the increased motor traffic through its trunk roads has destroyed many animals and necessitated the construction of fences to protect man and animal from each other. The use of insecticides extends beyond their effect on insects as many of these substances remain toxic and may kill the predators. These are but a few of the problems which must be considered if we are to keep our native species. Conversely, populations of some species must not become too large. The demographers have already found many difficulties and pitfalls in trying to predict future changes in the human population. It is therefore unlikely that there will be any greater accuracy in predictions of the future animal populations. The question that remains is whether the successful will become more successful and the unsuccessful extinct.

[1] Provisional distribution maps of British mammals. G. B. Corbett in Mammal Review, volume 1, pp. 95–142, 1971.

[2] A preliminary account of the mapping of the distribution of the freshwater fish of the British Isles. P. S. Maitland in Journal of Fish Biology, volume 1, pp. 45–58, 1969.

GLOSSARY

Aestivation. The dormant state of an animal during summer or a dry season.

Ammocoete. The larval stage of a lamprey, which may last for several years after hatching.

Amphipod. A small, laterally flattened crustacean like the sand hopper.

Bathypelagic. Living in the depths of the ocean, but not on the bottom.

Benthic. Living on the bottom below a body of water.

Brackish water. A mixture of salt and fresh water.

Canine tooth. A pointed tooth, found beside incisor teeth on each side of the upper and lower jaws of mammals, sometimes enlarged to form tusks. Also present in some fishes.

Carnivore. A flesh-eating animal, for example, a member of the cat family.

Cephalopod. A mollusc with a distinct head, like the cuttlefish, squid or octopus.

Chromatophore. A pigment-containing cell, which by contraction or relaxation provides a change in coloration.

Chromosome. A thread-like structure controlling the development of genetic characteristics, found in pairs in the nucleus of plant and animal cells. Each organism has a fixed number of them in each cell, half the total coming from each parent.

Claspers. Scroll-like organs, the modified parts of pelvic fins of male cartilaginous fish, which are inserted into the female during copulation.

Cloaca. In fish, amphibians, reptiles, and birds, a common terminal chamber into which genital, urinary, and intestinal canals open.

Coelenterate. An aquatic, invertebrate animal, usually radially symmetrical, that has a simple gut with a single opening, like the sea-anemone and *Hydra*.

Coitus. The act of sexual intercourse between the male and female of any species in which fertilization takes place within the parent body. Another term for this is copulation.

Continental shelf. The relatively shallow platform extending from the edge of a continent beneath the sea to a water depth of about 200 m.

Continental slope. A steep downward slope from the outer edge of the continental shelf to the depths of the sea.

Copepod. A small crustacean without a carapace, like *Calanus* and *Cyclops*.

Copulation. see coitus.

Crustacean. An invertebrate, segmented animal with jointed limbs, encased in a chitinous exoskeleton. Most are aquatic; the crab, prawn, and shrimp are examples.

Delayed implantation. A pause, sometimes very prolonged, that occurs in some mammals after fertilization, when the embryo lies dormant in the uterus before beginning to grow. First described by William Harvey in 1651 from his observations of roe deer; now known in the badger and the common seal too.

Demersal. Found on or near the bottom of a body of water.

Denticle. A small tooth-like protuberance or scale of dentine and enamel, usually scattered over the skin of elasmobranch fishes.

Dew claw. The vestiges of a toe above and behind the principal toes, as on the inside of a dog's paw.

Diastema. The space between teeth of various types found in the jaws of some mammals, for example, the horse or rabbit.

Dimorphism. Two distinct forms of structure, colour, or size among animals of the same species. Sexual dimorphism — forms differing in the male and female, as in the deer, where the male bears antlers.

Dorsal. Situated on the back, or on the side of an animal which is normally directed upwards.

Elasmobranch. A fish whose skeleton is made up of cartilage rather than bone.

Elver. An eel in the stage of development that follows immediately after metamorphosis.

Embryo. An organism in the early stages of development, before birth or hatching.

Enamel. The hard, glossy covering of a tooth.

Epifauna. Animals living on the ground.

Euphausid. A small, marine, planktonic crustacean with stalked eyes and light organs These animals are an important part of the 'krill', the food of baleen whales.

Fecund. Fertile, capable of producing offspring.

Feral. Existing in a wild state, or having escaped from domesticated state to live in the wild.

Fertilization. The union of two special cells, the gametes, ovum and sperm, for the initiation of the development of a new individual.

Foetus. The unborn offspring developing in the uterus. Term used only after young has come to show external and internal features resembling those of the adult form.

Fry. Young fish after completion of metamorphosis when adult fins have formed and scales have been laid down.

Gamete. A reproductive cell. Gametes from male (sperm) and female (ovum) join to start the development of a new individual.

Genital papilla. An external protuberance of the organs of reproduction.

Genus (pl. genera). A group of closely-related species.

Gestation period. The time from conception to birth in a viviparous animal.

Gonad. The organ of a male or female in which gametes, sperm or ova, develop.

Habitat. The natural surroundings of an animal.

Hand. A measurement, based on the width of the palm of a man's hand (102 mm or 4 inches), used to indicate the standing height of certain animals, particularly the horse and the deer.

Herbivore. An animal which feeds on plants.

Heat. In some animals, a time of sexual excitement, when the female is receptive to the male and able to conceive.

Hibernation. A dormant state during winter, when the animal becomes cooler and there is a slowing of bodily functions.

Home range. The area in which an animal normally lives and moves about.

Incisor tooth. One of the anterior teeth of the upper and lower jaw, used for cutting.

Indigenous. Native to a particular area.

Isopod. A small crustacean, usually with a dorso-ventrally flattened body, like the wood- or water-louse.

Lactation period. The time when the mammary glands are producing milk, from birth to weaning when the young are suckled by their mother.

Larva (pl. larvae). The form of some animals after hatching, not showing a close resemblance to the adult, for example, the young of the frog, newt, and many fish.

Lateral. Situated on the side of the body.

Leptocephalus larva. The thin-headed, leaf-like, transparent larva of the eel, found near the surface of the sea. This stage is followed by the metamorphosis to an elver.

Littoral zone. The region of the shore bounded by extreme high water and the lowest level of spring tides.

Metamorphosis. A change in body form that takes place during the life of certain animals, for example, the development from tadpole to frog.

Milt. The product of the male fish gonad containing sperm. It is ejected over eggs to fertilize them.

Molar tooth. A mammalian cheek tooth with an expanded crown and a complicated pattern of projecting cusps. The molar teeth of upper and lower jaws meet closely, in occlusion, to break up food.

Musk gland. A gland secreting a substance with a characteristic strong smell, found in the musk deer, badger, otter, and other animals.

Mysid. A small, shrimp-like crustacean.

Notochord. A rod-like structure, the chief support of the body of lower chordates and cyclostomes. In higher vertebrates, found only in the embryo.

Oestrus. The time of sexual activity, or heat, in the female. It occurs either rhythmically throughout the year, or in one or more seasons, when eggs are released from the ovary and the female is most receptive to the male.

Omnivore. An animal which is able to eat both plant and animal food.

Oocyte. The female germ cell in a mature state, containing half the adult number of chromosomes.

Operculum. An external flap covering the gills of bony fish.

Otolith. A mass of granules of calcium carbonate found in the inner ear of vertebrates and in the organ of balance, otocyst, of invertebrates.

Oviparity. A mode of reproduction in which the eggs are not retained in the body after maturation and fertilization.

Ovoviviparity. A mode of reproduction in which eggs are retained in the body of the female and hatch there.

Ovum (pl. ova). A female unfertilized reproductive cell (gamete or oocyte) which contains half the adult number of chromosomes and which, after fertilization by a sperm is capable of developing into a new individual.

Parturition. In mammals, the act of giving birth.

Pectoral fins. Paired fins of fish which lie behind the head, attached to the pectoral girdle.

Pelagic. Living freely in the open sea or the ocean.

Pelvic fins. Paired fins of fish on the lower surface of the body, attached to the pelvic girdle.

Placenta. A special organ that develops from the uterine wall and foetal membranes, to allow exchange of oxygen, food, and waste products between mother and foetus, while keeping their individual blood supplies separate.

Plankton. Minute plants (phytoplankton) and animals (zooplankton) that drift in salt and fresh water, forming the food of many of the larger animals.

Polychaete. An annelid worm, usually marine.

Post-larva. A young fish which has absorbed the yolk sac of its egg but has not yet developed adult fins.

Premolar tooth. One of a series of teeth that lie behind the canines on each side of both jaws of mammals. The crown frequently has a grinding surface.

Roe. A mass of eggs (also called ova or spawn) produced by the female fish.

Rut. A periodically recurring season of sexual excitement in certain male animals, for example, the deer or the goat.

Sexual maturity. The attainment of an age or size when the reproductive organs have matured so that ova and sperm produced by them are capable of fertilization.

Species. The basic unit of classification of animals and plants; a group of individuals that can breed among themselves but not with other species.

Spent fish. A female fish after spawning, usually in a weak and exhausted condition.

Sperm. A small, motile and usually flagellated male gamete, carrying half the adult number of chromosomes.

Spiracle. A gill-opening which has become modified. It lies behind the head of rays and many sharks and is large in bottom-living species to provide an inflow of clear water to gills, but small or absent in fast-swimming types of shark.

Sub-littoral zone. A region extending from the limit of the shore to the margin of the continental shelf.

Subspecies. A geographically definable, taxonomically distinct population within a species, though still capable of interbreeding with other members of the species. If it remains in isolation a subspecies may develop into a species.

Taxonomy. The discipline concerned with the identification, naming and classification of organisms.

Tubercle. A small, rounded protuberance on the surface of the body.

Uterine milk. A secretion produced by villi on the uterine wall to supply nutrient to the embryo.

Ventral. Situated on the front, or the side of an animal usually directed downwards.

Villus (pl. villi). A finger-like projection of tissue that increases the surface area of the organ in which it is situated.

Viviparity. A mode of reproduction in which the young are born alive, found in mammals, some reptiles, and a few fish. During gestation the placenta provides an intimate connection between mother and young.

Withers. The ridge between the shoulder blades of a horse or a deer.

Yolk sac placenta. A specially modified portion of the yolk sac found in some species of cartilaginous fish. It comes into contact with the uterine wall through which oxygen and nutrient material pass to the embryo and waste products are removed.

UNITS OF MEASUREMENT

Metric	Abbreviation	Imperial	(approx.)
WEIGHT			
1 gramme	g	0·0353 ounces	$\frac{1}{30}$ ounce
1 kilogramme	kg	2·2046 pounds	$2\frac{1}{4}$ pounds
LENGTH			
1 millimetre	mm	0·0394 inches	$\frac{1}{25}$ inch
1 centimetre	cm	0·3937 inches	$\frac{2}{5}$ inch
1 metre	m	3·281 feet	$3\frac{1}{4}$ feet
1 kilometre	km	0·621 miles	$\frac{5}{8}$ mile
AREA			
1 hectare	ha	2·471 acres	$2\frac{1}{2}$ acres

SOURCES OF FURTHER INFORMATION

General
KERRICH, G. J., MEIKLE, R. D. & TEBBLE, N. *Bibliography of key works for the identification of the British fauna and flora*. 3rd edn. 1967. Systematics Association, London.
ROMER, A. S., *Vertebrate paleontology*. 1966. University of Chicago Press.
ROMER, A. S., *The vertebrate body*. 4th edn. 1970. Saunders, London.
The living world of animals. 1970. Reader's Digest Association, London.
YOUNG, J. Z., *The life of vertebrates*. 2nd edn. 1962. Clarendon Press, Oxford.

Cyclostomes
HARDISTY, M. W. & POTTER, I. C. *The Biology of lampreys*. I, 1971. Academic Press, London.

Fish, cartilaginous and bony
BAGENAL, T. B., *The fauna of the Clyde Sea area: Fishes*. 1965. Scottish Marine Biological Association, Millport.
BRUCE, J. R., COLMAN, J. S. & JONES, N. S., *Marine fauna of the Isle of Man and its surrounding seas*. 1963. Liverpool Marine Biological Committee Memoir No. 36. Liverpool University Press.
HARDY, A., *The open sea: Its natural history, Part II Fish and Fisheries*. 1959. Collins, London.
JENKINS, J. T., *The fishes of the British Isles both fresh water and salt*. Reprinted 1958. Warne, London.
KENNEDY, M., *The sea angler's fishes*. 2nd edn. 1969. Paul, London.
MAITLAND, P., *Key for the identification of the freshwater fishes of the British Isles, with notes on their ecology and distribution*. 1972. Freshwater Biological Association.
MARINE BIOLOGICAL ASSOCIATION, 1957, *Plymouth Marine Fauna*.
MEEK, A., *The migrations of fish*. 1916. Arnold. London.
MILLS, D., *Salmon and trout: A resource, its ecology, conservation and management*. 1971. Oliver & Boyd, Edinburgh.
MUUS, B. J. & DAHLSTRØM, P. *Freshwater fish of Britain and Europe*. 1971. Collins. London.
VARLEY, M. E. *British freshwater fishes: Factors affecting their distribution*. 1967. Fishing News, London.
WENT, A. E. J. & KENNEDY, M., *List of Irish fishes*. 2nd edn. 1969. The Stationery Office, Dublin.

Amphibia and reptiles
APPLEBY, L. G.. *British snakes*. 1971. Baker, London.
BELLAIRS, A. d'A., *Reptiles*. 3rd edn. 1970. Hutchinson, London.
SIMMS, C., *Lives of British lizards*. Goose & Son, Norwich.
SMITH, M., *The British amphibians and reptiles*. 4th edn. 1969. Collins, London.

Mammals
CORBET, G. B., *The identification of British mammals*. 1964. British Museum (Natural History), London.
LAWRENCE, M. J. & BROWN, R. W., *Mammals of Britain: Their tracks, trails and signs*. 1967. Blandford, London.
MATTHEWS, L. H. *British mammals*. 2nd edn. 1968. Collins, London.
MELLANBY, K., *The mole*. 1971. Collins, London.
NEAL, E. *The badger*. 1948. Collins, London.
OMMANNEY, F. D., *Lost leviathan*. 1971. Hutchinson, London.
PAGE, F. J. T., *Field guide to British deer*. 2nd edn. 1971. Blackwell, Oxford.
SHORTEN, M., *Squirrels*. 1954. Collins, London.
SOUTHERN, H. N. (editor) *The handbook of British mammals*. 1964. Blackwell, Oxford.

Societies and Journals
The Council for Nature, Zoological Gardens, Regent's Park, London NW1 4RY is the representative body of the natural history movement of the United Kingdom from whom the names and addresses of the many societies can be obtained. Most of these societies publish their own journal.

INDEX

Abramis 194; *brama* 96
Acanthopterygii 194
Acerina cernua 94
Acipenser 193; *ruthenus, sturio* 88
Acipenseriformes 88, 193
Actinopterygii 193
Adaptation 202; for bottom-living
  10–18, 40, 44–54; for diving 182;
  for flight 154; for jumping 116;
  for swimming 116, 174, 176, 182
Adder 124, **125**, 202
Adipose fin 88, 102, 193, 194
Age, of plaice 44; of trout 108
Agnatha viii, 192
Agonidae 80
Agonus 195; *cataphractus* 80
Alar spine **vii**
Albacore 26
Alburnus 194; *alburnus* 98
Alevins 104–108
Alligator 196
Allis shad 22, **23**
Alopias 192; *vulpinus* 4
Alosa 193; *alosa, fallax* 22
Alytes 196; *obstetricans* 118
Ambergris 188
Amberjack 30, **31**
American Mink 168, **169**
Ammocoete larva viii, **1**
Ammodytes 195; *immaculatus,
  lanceolatus, marinus, tobianus* 60
Ammodytoidei 60, 195
Amphibia vii, 114–118, 196, 201;
  fossils of 198
Amplexus 116, 118
Anadromous fish 88
Anal fin **vii**
Anapsida 196
Anchovy 20
Angelfish 10
Angler Fish 40, **41**, 194
Anguilla 193; *anguilla* 86
Anguillidae 86
Anguilliformes 84–86, 193
Anguis 196; *fragilis* 122
Antelope 197
Antlers **vii**, 142–149
Anura 116–118, 196
Anus vi
Aphia 195; *minuta* 56
Aphrodisiac 120
Aphya minuta 56

Apletodon 194; *microcephalus* 78, **79**
Apodemus 198; *flavicollis,
  sylvaticus* 132
Aquatic mammals 182–190
Archosauria 196
Arctic Hare 140
Arctic Ocean 199, **200**
Argentina 193; *silus* 24;
  sphyraena 22
Argentine(s) 22, 24, 193
Argentinoidei 22, 24, 193
Arnoglossus 195; *imperialis, laterna,
  thori* 52
Artificial pearls 98
Artiodactyla 142–150, 197
Arvicola 198; *amphibius,
  terrestris* 130
Aspitrigla 195; *cuculus, obscura* 68
Atherina 194; *presbyter* 74
Atheriniformes 32, 74, 194
Atherinomorpha 194
Atlantic Ocean 199, **200**
Attachment mechanisms viii, 42, 56,
  78, 80, 120
Auxis 195; *thazard* 24
Aves 196

Backbone vi
Badger 164, **165**, 197
Balaena 197; *glacialis* 190
Balaenidae 190
Balaenoptera 197; *acutorostrata,
  borealis, musculus, physalus* 190
Balaenopteridae 190
Baleen plates 190, 197
Baleen whales 190, 197, 199
Balistes 196; *capriscus,
  carolinensis* 42
Balistoidei 196
Ballan Wrasse 62, **63**
Bank Vole 130, **131**
Barbastella 197; *barbastellus* 160
Barbastelle 160, **161**
Barbel 98, **99**
Barbel(s) 194; of catfish 92;
  of cod and family 34, 36, 92;
  of loach 92, 100; of red mullet 30;
  of rocklings 76; of sturgeon 88,
  193
Barbus 194; *barbus* 98
Bare buck 146

Barking Deer 148
Basking Shark 4, **5**, 199
Bass 58, **59**, 200
Bat(s) vii, 154–162, 197, 202
Bathypelagic 199
Batoidea 10–18, 193
Beak, of dolphin **vii**, 182–184;
  of fish 42; of turtle 120;
  of whale 182, 188
Bear 201
Bearded Seal 180, **181**
Bechstein's Bat 158, **159**
Bed of salmon 104
Belone 194; *bellone* 32
Belted Bonito 26
Beluga 186
Benthic fish: angler fish 40;
  flatfish 44–54; gobies 56, 82;
  rays 10, 12; sand eels 60;
  sharks 6–10; weevers 58
Bib 36
Billy goat 150, **151**
Biological Records Centre 202
Bird(s) vi, 196; fossils of 198
Birth, of bat 162; of deer 142;
  of dolphin 184; of porpoise 182;
  of seal 178
Biscayan Right Whale 190
Bitch: weasel 170; otter 174
Bitterling 112, **113**
Black Bream 28, **29**
Black Goby 82, **83**
Black Rat 134, **135**
Black-mouthed Dogfish 6, **7**, 200
Bleak 98, **99**, 201
Blennies 72, 195
Blennioidei 72, 74, 195
Blennius 195; *galerita, gattorugine,
  montagui, ocellaris, pholis* 72
Blicca 194; *bjoerkna* 96
Bloater 20
Blonde Ray 12, **13**
Blood vi, 126
Blow-hole **vii**, 182
Blowing, of whales 188–**191**
Blubber 176, 180, 182
Blue Hare 140, **141**
Blue Shark 4, **5**
Blue Skate 14
Blue Whale 190, **191**
Blue Whiting 36
Bluefin Tuna 26

Boar: badger 164
Body temperature vi, 26, 196, 202
Bogue 28, **29**
Bone vi
Bonitos 26
Bony, armour 40, 194; fish vi, **vii**,
   20–112; scutes 88, 193;
   skeleton 20, 88, 114, 193
Boops 195; *boops* 28
Borer viii
Bos 197; *taurus* 150
Bothidae 52, 54
Bottle-nosed Dolphin 184, **185**
Bottle-nosed Ray 14, **15**
Bottle-nosed Whale 188, **189**
Bottom-living, animals 200:
   bony fish 40; flatfish 44–54;
   gobies 56, 82; rays 10–12;
   sand eels 60; sharks 6–10;
   weevers 58
Bovidae 150
Bowfin 193
Brackish water 200
Brain vi, 126, 192, 196
Brama 195; *brama* 30
Bramble Shark 6, **7**
Bramidae 30
Bream 28, 96, **97**, 201
Breeding behaviour 90
Breeding season, of deer 142–148;
   of dormouse 132; of hare 140;
   of hedgehog 126; of mouse 132,
   134; of rabbit 138; of rat 134;
   of shrew 128; of squirrel 136;
   of vole 130; of weasel 170;
   of whale 188
Brill 54, **55**
British Isles **200**, 202
Broad-nosed Pipefish 64, **65**
Brock 164
Brood pouch 64
Brook Lamprey viii, **1**
Brosme 194; *brosme* 36
Brown Hare 140, **141**, 202
Brown Rat 134, **135**, 202
Brown Trout 108, **109**
Buck: deer 144–149; rabbit 136
Buckling 20
Buenia 195; *jeffreysii* 56
Bufo 196; *bufo, calamita, viridis* 118
Bufonidae 118
Buglossidium 195; *luteum* 50
Bull: Chillingham 150, **151**;
   common seal 176; grey seal 178,
   179; killer whale 186; reindeer
   148, **149**

Bull Huss 8, **9**
Bullhead(s) 66, **67**, 100, **101**, 194, 201
Buoyancy 10
Burbot 92, **93**
Butterfish 74, **75**
Butterfly Blenny 72, **73**

Calanus 20, **21**
Calcar **vii**
Calf: bottle-nosed dolphin 184, **185**;
   Chillingham 150, **151**;
   common rorqual 190, **191**;
   pilot whale 186; red deer 142, **143**;
   sika deer 146, **147**
Callionymoidei 70, 195
Callionymus 195; *lyra, maculatus,
   reticulatus* 70
Camouflage, of deer 144;
   of dragonet 70, **71**; of frog 116;
   of plaice 44, **45**; of toad 118
Canidae 166
Canine teeth **vii**, 142, 172
Capra 197; *hircus* 150
Capreolus 197; *capreolus* 144
Capromyidae 130
Carangidae 30
Carapace 120
Carassius 194; *auratus, carassius* 112
Caretta 196; *caretta* 120
Carnassial teeth 172
Carnivora 164–174, 197
Carnivore(s) 114–124, 164–190, 197
Carp 112, **113**, 194, 201
Cartilage vi, 192
Cartilaginous, fish **vii**, 2–18, 88;
   skeleton vi, viii, 192
Cat **vii**, 172, 197
Catadromous fish 86
Catfish 92, **93**, 194
Cattle 150, 197
Caudal fin **vii**
Caviar 88
Caviomorpha 198
Centrarchidae 112
Centrolabrus 195; *exoletus* 62
Cepola 195; *rubescens* 30
Cepolidae 30
Cervidae 142–148
Cervus 197; *elaphus* 142; *nippon* 146
Cetacea 182–190, 197
Cetorhinus 192; *maximus* 4
Channel Islands **200**
Chaparrudo 195; *flavescens* 82
Char 110, **111**, 201
Chelonia 120, 196

Chillingham Cattle 150, **151**
Chimaera 193; *monstrosa* 18
Chimaeriformes 193
Chinese Muntjac 148, **149**
Chinese Water Deer 148, **149**
Chirolophis 195; *ascanii* 72
Chiroptera 154–162, 197
Chlamydoselachus 192; *anguineus* 2
Chondrichthyes 2, 192
Chondrostei 88, 193
Chromatophores 44, 66, 78, 114, 116
Chub 98, **99**, 201
Ciliata 194; *mustela,
   septentrionalis* 76
Claspers **vii**, 2, 6, 8, 18
Class 192
Clethrionomys 198; *glareolus* 130
Clingfish 194
Clione **21**
Cloaca, of newt 114; of shark 8
Clupea 193; *harengus* 20
Clupeidae 20, 22
Clupeiformes 193
Clupeomorpha 20, 193
Coalfish 38, **39**
Coat, change of, in deer 144;
   in hare 140, **141**; in reindeer 148;
   in seal 176–180; in stoat 170, **171**;
   in squirrel 136
Cobitidae 92, 100
Cobitis 194; *taenia* 92
Cod 34, **35**, 194
Coelacanth 196
Coley 38
Colias Mackerel 24, **25**
Colour changes, of frog 116;
   of plaice 44, **45**; of sucker 78;
   of toad 118
Colubridae 124
Comber 58
Common Dogfish 8
Common Dolphin 182, **183**
Common Dragonet 70, **71**
Common Fin Whale 190
Common Frog **vi**, 116, **117**
Common Goby 82, **83**
Common Hare 140
Common Lizard 122
Common Long-eared Bat 160, **161**
Common Porpoise 182, **183**
Common Rorqual 190, **191**
Common Sea Bream 28
Common Seal 176, **177**, 200
Common Shrew 128, **129**, 202
Common Skate 14, **15**
Common Toad 118, **119**

Common Topknot 52, **53**
Conger 193; *conger* 84
Conger Eel 84, **85**
Congridae 84
Connemara Pony 152
Connemara Sucker 78, **79**
Continental shelf **199**, 200;
  fish of 22, 26, 28, 40, 86, 88
Continental slope **199**, 200;
  fish of 6, 18, 84
Coprophagy 138
Copulation, in fish 8, **9**;
  in mammals 140, 154, 156, 162,
  168, 170, 176–178, 184;
  in reptiles 120–122
Corbet, G. B. 192
Coregonus 193; *albula, clupeoides,
  pollan* 110
Coris 195; *julis* 62
Corkwing Wrasse 62, **63**
Cornish Sucker 78, **79**
Coronella 196; *austriaca* 124
Coryphoblennius 195; *galerita* 72
Cottidae 66, 100
Cottus 195; *bubalis* 66; *gobio* 100;
  lilljeborgi, scorpius 66
Couch 174
Couch's Sea Bream 28, **29**
Coursing 140
Courtship behaviour, in amphibia
  114–118; in fish 62, 66, 70, 90,
  104, 108; in mammals 126, 138–
  146, 164, 176, 182; in reptiles 124
Courtship dress, of dragonet 70, **71**;
  of leopard spotted goby 82;
  of minnow 100; of newt 114;
  of salmon 104, **105**; of sea
  scorpion 66; of shanny 72
Cow: Chillingham 150, **151**;
  common seal 176; grey seal 178,
  179; killer whale 186;
  reindeer 148, **149**
Coypu 130, **131**, 198, 201, 202, 203
Crampfish 10
Cranium 192
Crenilabrus 195; *melops* 62
Crenimugil 195; *labrosus* 74
Cretaceous Period 193, 198
Crocidura 197; *russula,
  suaveolens* 128
Crocodile 196
Crucian Carp 112, **113**
Crustacea 20, **21**, 199
Crystal Goby 56, **57**
Crystallogobius 195; *linearis,
  nilssoni* 56

Ctenolabrus 195; *rupestris* 62
Cubs, of fox 166, **167**; of otter 174,
  175
Cuckoo Ray 16, **17**
Cuckoo Wrasse 62, **63**
Cuvier's (Beaked) Whale 188, **189**
Cyclopteridae 80
Cyclopterus 195; *lumpus* 80
Cyclostomata viii, 192
Cyclostomes **vii**, viii
Cyprinidae 94–102, 112
Cyprinids 94–102
Cypriniformes 92–94, 100–102, 194
Cyprinus 194; *carpio* 112
Cystophora 197; *cristata* 180

Dab 48, **49**
Dace 102, **103**, 201
Dalatias 192; *licha* 8
Dales Pony 152
Dama 197; *dama* 146
Darker Three-bearded Rockling 76,
  77
Darkie Charlie 8, **9**, 200
Dartmoor Pony 152, **153**
Dasyatis 193; *pastinaca* 18
Daubenton's Bat 158, **159**
Deer **vii**, 142–148, 197
Defensive weapons, of dogfish 6;
  of ray 10, 18; of weever 58
Delayed fertilization 154, 162
Delayed implantation, of badger 164;
  of mink 168; of otter 174;
  of pine marten 170; of roe deer
  144; of seal 176–180; of stoat 170
Delphinapterus 197; *leucas* 186
Delphinus 197; *delphis* 182
Den, of fox 166; of otter 174
Dentex 28
Dentex 195; *dentex* 28
Denticle(s), dermal **vi**, 2, 10
Dermochelys 196; *coriacea* 120
Devil Fish 18
Devonian Period 192, 198
Diastema 130
Dicentrarchus 195; *labrax* 58
Diet vi, 202
Digestion 142, 152
Digestive system vi
Diminutive Goby 56, **57**
Diplecogaster 194; *bimaculata* 78
Disc, for attachment 42; of ray **vii**,
  10–16; of skate 14
Distribution map of animals 202
Doe: Chinese muntjac 148, **149**;

Chinese water deer 148, **149**;
  fallow deer 146, **147**; Indian
  muntjac 148; rabbit 138; roe
  deer 144, **145**
Dog: fox 166; otter 174; weasel 170
Dogfish vi, **vii**, 6, 8, 200
Dolphin(s) **vii**, 182–**187**, 197
Dormouse 132, **133**, 198, 202
Dorsal fin **vii**
Dragonet(s) 70, 195
Drey, squirrel's 136, **137**
Duck-billed Platypus 196
Dusky Perch 58

Eagle Ray 18, **19**
Earth, fox's 166
Echinorhinus 192; *brucus* 6
Echolocation, bat 154; dolphin 182
Eckström's Topknot 52, **53**
Edible Dormouse 132, **133**
Edible Frog 116, **117**
Eel(s) 84–**87**, 193, 200, 201
Egg(s) vi; adhesive **21**, **79**, **81**, 96;
  attachment filaments of 8, 10, 32,
  78; demersal 22, 40, 56, 66;
  hatching of **45**
Egg(s), of butterfish 74, **75**;
  of Cornish sucker 78, **79**;
  of dogfish 8, **9**; of eel 86, **87**;
  of frog 116, **117**; of hagfish viii, **1**;
  of lumpsucker 80, **81**; of newt 114,
  115; of plaice 44, **45**; of skate 14,
  15; of stickleback 90, **91**; of toad
  118, **119**; of turtle 120
Egg-laying mammal 196
Elasmobranchii 192
Electric organ 10
Electric Ray 10, **11**
Elephant 201
Elopomorpha 193
Elvers 86, **87**
Enchelyopus cimbrius 76
England **200**, 201
English Channel **200**
Engraulis 193; *encrasicolus* 20
Entelurus 194; *aequoreus* 64
Eocene Period 126
Epinephalus 195; *guaza* 58
Epipelagic 199
Eptesicus 197; *serotinus* 162
Equidae 152
Equus 197; *caballus* 152
Erignathus 197; *barbatus* 180
Erinaceus 197; *europaeus* 126
Ermine 170

Esocidae 110
Esocoidei 110, 193
Esox 193; *lucius* 110
Estuary 200
Euphrosyne Dolphin 182
European Toad 118
European Tree-Frog 118
Eutheria 197
Euthynnus 195; *alletteratus* 26
Eutrigla 195; *gurnardus* 68
Even-toed ungulate 197
Ewe 150, **151**
Ewing's Frog 118
Exmoor Pony 152
Exocoetus 194; *volitans* 32
External gills 114–118
Extinct mammals 201
Eye **vii**

Fallow Deer 146, **147**
False Killer 186, **187**
Family 192
Faroe Islands 199, **200**
Fat Dormouse 132
Fawn: fallow deer 146, **147**
Felidae 172
Felis 197; *sylvestris* 172
Fell Pony 152
Ferret 168, **169**
Fertilization, external viii, 20–112,
    116; internal 76, 114, 120–190, 196
Field Rat 134
Field Vole 130, **131**
Fifteen-spined Stickleback 90, **91**
File-fish 42, 196
Filter feeder, basking shark 4;
    whale 190
Fin(s) vi, **vii**, 192, 193
Fingerling 104, 108
Finlets 24, 195
Fire Flair 18
Fish vi, **vii**; bony 20–112;
    bottom-living 6–12, 40, 44–60, 82;
    cartilaginous 2–18; coastal 56–82;
    colour 200; deep-sea 22, 40;
    flat 44–54, 195, 200; flying 32,
    199; fossils of 198; lake 201;
    largest 4; larvae 199;
    migratory 22, 46, 86, 88, 90, 200;
    rearing 44, 54, 108; shape 200;
    smallest 90; sound produced by
    34, 66
Fissipedia 197
Five-bearded Rockling 76, **77**
Flake 6

Flatfish 44–54, 195, 200
Flight, vi; in fish 32, **33**;
    in mammals 154–162, 197, 202
Flipper(s) **vii**, 176, 182, 197
Flounder 46, **47**, 200
Flying Fish 32, **33**, 194, 199
Foal: New Forest pony 152, **153**
Forked Hake 36
Form, brown hare's 140, **141**
Fossils 192, 198; of vertebrates vi,
    192, 193
Four-bearded Rockling 76, **77**
Fox 166, **167**, 197
French Sole 50
Freshwater fish 90–112
Freshwater vertebrates 201
Fries's Goby 56, **57**
Frigate Mackerel 24
Frilled Shark 2, **3**
Frog(s) **vi**, **vii**, 116, 196;
    spawn of 116, **117**
Frogfish 40
Fry, of salmon 104; of trout 108

Gadidae 34, 36, 38, 76, 92
Gadiformes 34–38, 76, 92, 194
Gadus 194; *esmarkii*, *luscus*,
    minutus 36; *morhua* 34;
    pollachius 38; *poutassou* 36;
    virens 38
Gaidropsarus 194; *mediterraneus*,
    vulgaris 76
Galeorhinus 192; *galeus* 8
Galeus 192; *melastomus* 6
Garfish 32, **33**
Garpike 193
Gasterosteidae 90
Gasterosteiformes 64, 90, 194
Gasterosteus 194; *aculeatus* 90
Genus 192
Geological, Periods **198**; structure
    201; time-scale **198**
Germo alalunga 26
Gestation period, of snake 124;
    of mammals 126–190
Giant Goby 56
Gill vi, **vii**, 192, 193; chamber 88;
    cover **vii**; openings **vii**, 2, 10, 20,
    84
Gills, external 114–119
Gilt-head 28, **29**
Girdle, pectoral vi; pelvic vi
Glacial Period 201
Gland, musk 130, 164, 197;
    sub-orbital **vii**

Gliridae 132
Glis 198; *glis* 132
Globicephala 197; *melaena* 186
Glyptocephalus 195; *cynoglossus* 46
Goat 150, **151**, 197
Gobies, 197; coastal 82, 200;
    off-shore 56; pelagic 56
Gobiesociformes 78, 194
Gobio 194: *gobio* 98
Gobioidei 56, 82, 195
Gobius 195; *cobitus* 56; *cruenatus* 82;
    fagei 56; *flavescens*, *forsteri*,
    microps, *minutus*, *niger*,
    pagenellus, *pictus*, *ruthensparri* 82
Golden Grey Mullet 74, **75**
Goldfish 112, **113**
Goldsinny 62, **63**
Goureen 22
Grampus 197; *griseus* 184
Grass Snake 124, **125**
Grayling 102, **103**
Great Buck 146
Great Crested Newt 114
Great Silver Smelt 24
Greater Argentine 24
Greater Fork-beard 36, **37**
Greater Horseshoe Bat 154, **155**
Greater Pipe-fish 64, **65**
Greater Sand Eel 60, **61**
Greater Spotted Dogfish 8
Greater Weever 58, **59**
Green Lizard 122
Greenbone 32
Greenland Shark 4, **5**
Grey Gurnard 68, **69**
Grey Long-eared Bat 160, **161**
Grey Mullet 74, **75**, 200
Grey Seal 178, **179**
Grey Skate 14
Grey Squirrel 136, **137**, 202, 203
Guanine 104
Guinea Pig 198
Gudgeon 98, **99**, 201
Guernsey Vole 130
Gunnel 74
Gurnard 68, **69**, 194
Gymnammodytes 195;
    semisquamatus 60
Gymnocephalus 195; *cernua* 94

Habitat vi, 202, 203
Haddock 34, **35**
Hagfish viii, **1**, 192
Hake 38, **39**
Halibut 46, **47**, 195

Halichoerus 197; *grypus* 178
Hammerhead 8, **9**
Harbour Seal 176
Hare(s) 140, **141**, 198
Harp Seal 180, **181**
Harvest Mouse 132, **133**
Heart vi
Hebrides **200**
Hedgehog 126, **127**, 197
Hemipenes 122
Herbivore(s) 130–132, 136–152, 197
Herring vi, **vii**, 20, **21**, 193, 199
Hexanchus 192; *griseus* 2
Hibernation, of bat 154–162;
    of dormouse 132; of frog 116;
    of hedgehog 126; of lizard 122;
    of snake 124; of toad 118
Highland Pony 152
Highlands 201
Hind: red deer 142, **143**;
    sika deer 146, **147**
Hippocampus 194; *ramulosus* 64
Hippoglossoides 195; *limandoides,*
    platessoides 48
Hippoglossus 195; *hippoglossus* 46
Hippopotamus 197, 201
Holocephali 18, 193
Holostei 193
Holt, otter's 174
Homelyn 12
Hooded Seal 180, **181**
Hoof 142
Hook, of salmon 104
Horned Ray 18
Horns 150
Horse 152, 197
Horse Mackerel 30
Horseshoe Bats 154, **155**
House Mouse 134, **135**, 202
House Rat 134
Humantin 6, **7**
Humpback Whale 190
Hybridization 96
Hydropotes 197; *inermis* 148
Hyla 196; *arborea, ewingii* 118
Hyperia **21**
Hyperoodon 197; *ampullatus* 188
Hyperoplus 195; *immaculatus* 60, **61**;
    lanceolatus 60

Ice Age 201
Ide 112
Idus idus 112
Iljinia pictus 82
Incisor teeth, of bony fish 28;

of horses 152; of lagomorphs 138;
    of rodents 130, 136
Indian Muntjac 148, **149**
Insecticide 203
Insectivora 126, 128, 197
Insectivores 126–128, 154–162
Interdigital membrane **vii**
Interfemoral membrane 156
Interglacial Period 201
Intestine vi
Introduced species 201, 203
Intromittent organ 8
Invertebrates, fossils of 198
Ireland **200**, 202
Irish Sea **200**
Isinglass 88
Isurus 192; *oxyrhinchus* 2
Ivory 180

Japanese Deer 146
Jaws vi, **vii**, 2, 42, 120, 192, 193, 194
Jeffrey's Goby 56, **57**
John Dory 40, **41**, 194

Katsuwonus 195; *pelamis* 26
Kelt 104
Kemp's Ridley 120
Kid: goat 150, **151**; roe deer 144,
    145
Killarney Shad 22
Killer Whale 176, 186, **187**
Kingdom 192
Kipper 20
Kitten: polecat 168; stoat 170, **171**;
    weasel 170; wild cat 172
Krill 190
Kype 104

Labroidei 62, 195
Labrus 195; *bergylta, mixtus,*
    ossifagus 62
Lacerta 196; *agilis, muralis, viridis,*
    vivipara 122
Lacertidae 122
Lacertilia 196
Lagenorhynchus 197; *acutus,*
    albirostris 184
Lagocephalus 196; *lagocephalus* 42
Lagomorpha 138, 140, 198
Lagomorphs 138–140
Lake-dwelling animals 201
Lamb 150, **151**
Lamna 192; *nasus* 2

Lampern viii, **1**
Lampetra 192; *fluviatilis, planeri* viii
Lamprey **vii**, viii, 192
Lampridiformes 40, 194
Lampris 194; *guttatus, luna* 40
Land carnivores 164–**175**;
    mammals 202; vertebrates 202
Large-mouthed Black Bass 112
Largest, animal 190; baleen whale
    190; dolphin 186; fish 4;
    toothed whale 188
Larva(e), of eel 86, **87**; of fish 199;
    of frog 116, **117**; of lamprey viii,
    1; of newt 114, **115**; of plaice 44,
    45; of toad 118, **119**
Lateral line **vii**, 32, 46
Latimeria 196
Leathery Turtle 120, **121**
Lebetus 195; *orca* 56
Leisler's Bat 162, **163**
Lemon 48
Lemon Sole 48, **49**
Leopard Spotted Goby 82, **83**
Lepadogaster 194; *bimaculata,*
    candollei, couchii, gouanii,
    lepadogaster, microcephalus 78
Lepidochelys 196; *kempii* 120
Lepidorhombus 195; *whiff-iagonis* 54
Lepidosauria 196
Leporidae 138, 140
Leptocephalus brevirostris 86
Leptocephalus larva 84–**87**, 193
Lepus 198; *capensis, europaeus,*
    timidus 140
Lesser Argentine 22, **23**
Lesser Fork-beard 36, **37**
Lesser Horseshoe Bat 154, **155**
Lesser Noctule 162
Lesser Pipe-fish 64
Lesser Rorqual 190, **191**
Lesser Sand Eel 60, **61**
Lesser Weever 58, **59**
Lesser White-toothed Shrew 128,
    129
Lesser-spotted Dogfish 8, **9**
Lesueurigobius 195; *friesii* 56
Leuciscus 194; *cephalus* 98; *idus* 112;
    leuciscus 102
Leverets 140, **141**
Light organ 22, 193
Limacina 20, **21**
Limanda 195; *limanda* 48
Limbs vi, 196
Ling 38, **39**
Linnaeus 192
Liparidae 80

Liparis 195; *liparis, montagui* 80
Lissamphibia 196
Liver vi, 10
Liza 195; *auratus, ramada* 74
Lizard 120–122, 196, 202
Locomotion, animal vi, 202
Loggerhead Turtle 120, **121**
Long Rough Dab 48, **49**
Long-finned Gurnard 67
Long-finned Tunny 26, **27**
Long-nosed Skate 14, **15**
Long-snouted rays 14, 16
Long-snouted skates 14
Long-spined Sea Scorpion 66, **67**
Lophiiformes 40, 194
Lophius 194; *piscatorius* 40
Lota 194; *lota* 36, 92
Lowlands 201
Lucioperca lucioperca 94
Luminescent organs 22, 193
Lumpenus 195; *lumpretaeformis* 72
Lumpsucker 80, **81**, 200
Lungfish 196
Lungs vi, 196
Lure 40, 194
Lutra 197; *lutra* 174

Mackerel 24, **25**, 195
'Mackerel guide' 32
Mackerel midge 76
Mackerel Shark 2, 199
Macropus 197; *rufogriseus* 197
Mail-cheeked fish 40, 68, 80, 100, 194
Mako Shark 2, **3**
Mammalia 126, 196
Mammals vi, **vii**, 126–190, 201, 202; fossils of 198
Mammary glands 126
Mammoth 201
Man, fossils of 198
Manta rays 18
Marbled Electric Ray 10, **11**
Marbled Tunny 26
Marine, carnivores 197; vertebrates 199
Marsh Frog 116, **117**
Marsupialia 197
Martes 197; *martes* 170
Maternal behaviour, of deer 142; of dolphin 184; of fox 166; of hedgehog 126; of rabbit 138; of seal 178–180
Mating behaviour, of badger 164;

of deer 142–146; of dogfish 8, **9**; of hare 140; of hedgehog 126; of porpoise 182; of rabbit 138; of seal 176; of skate 14
Maurolicus 193; *muelleri* 22
Meganyctiphanes **21**
Megaptera 197; *novaeangliae* 190
Megrim 54, **55**
Melanogrammus 194; *aeglefinus* 34
Meles 197; *meles* 164
Membrane **vii**
Merlangius 194; *merlangus* 34
Merlucciidae 38
Merluccius 194; *merluccius* 38
Mermaid's purse 8, **9**, 14, **15**
Mesopelagic 199
Mesoplodon 197; *bidens, mirus* 188
Mesozoic Period 193, 198
Metamorphosis, of atherinomorph 32; of eel 84–**87**; of flatfish 44–54; of frog 116; of herring family 20–22; of lamprey viii, **1**; of newt 114; of rockling 76; of sucker 78
Metatheria 197
Mice 132, 134, 197, 198
Microchirus 195; *boscanion, variegatus* 50
Micromesistius 194; *poutassou* 36
Micromys 198; *minutus* 132
Micropterus 195; *salmoides* 112
Microstomus 195; *kitt* 48
Microtus 198; *agrestis, arvalis, arvalis orcadensis, arvalis sarnius* 130
Midwife Toad 118
Migration, of bat 156; of bony fish 20–26, 30–38, 58–60, 74, 86–90, 100, 104–108, 200; of flatfish 48–54; of lampreys viii; of land animals 201; of rays 12; of sharks 2, 6, 8; of toads 118; of whales 190
Miller's Thumb 100
Mink 168
Minnow 100, **101**, 201
Mobula 193; *mobular* 18
Mobulidae 18
Modification, of pectoral fin 68; of pelvic fin 6, 8, 78
Mola 196; *mola* 42
Molar teeth, of fish 28; of horse 152
Mole 126, **127**, 197, 202
Molehills 126, **127**
Molluscs 199
Molva 194; *molva* 38

Monkfish 10, **11**
Monodon 197; *monoceros* 186
Monodontidae 186
Monotreme 196
Montagu's Blenny 72, **73**
Montagu's Sea Snail 80, **81**
Moonfish 40, 194
Moray Eel 84, **85**
Mormyrus 193
Morone 195; *labrax* 58
Mouse-eared Bat 156, **157**
Mouth vi, **vii**
Mugil 195; *auratus, capito, labrosus* 74
Mugiloidei 74, 195
Mullet(s) 74, 195, 200
Mullidae 30
Mullus 195; *surmuletus* 30
Muntiacus 197; *muntjak, reevesi* 148
Muntjacs 148, **149**
Muraena 193; *helena* 84
Muraenidae 84
Muridae 130–134
Mus 198; *mus muralis* 202; *musculus* 134
Muscardinus 198; *avellanarius* 132
Muscles vi
Musk glands 197; of badger 164; of vole 130
Musk Rat 130, **131**
Musk Shrew 128
Mussel **45**
Mustela 197; *erminea, nivalis* 170; *putorius, putorius furo, vison* 168
Mustelidae 164, 168, 170, 174
Mustelus 192; *asterias* 8; *mustelus* 6
Myliobatidae 18
Myliobatis 193; *aquila* 18
Myocastor 198; *coypus* 130
Myomorpha 198
Myotis 197; *bechsteini, daubentoni* 158; *myotis, mystacinus, nattereri* 156
Myoxocephalus 195; *scorpius* 66
Mysteceti 182, 190, 197
Myxine 192; *glutinosa* viii
Myxomatosis 138, 166, 203

Names, common, scientific vi
Nanny goat 150, **151**
Narwhal 186
Nathusius's Pipistrelle 158, **159**, 202
Natrix 196; *natrix* 124
Natterer's Bat 156, **157**
Natterjack 118, **119**

Naucrates 195; *ductor* 30
Neomys 197; *fodiens* 128
Nerophis 194; *lumbriciformis, ophidion* 64
Nervous system vi, 192
Nest, of brown hare 140, **141**; of bullhead 100; of cat 172; of char 110; of cuckoo wrasse 62; of dormouse 132, **133**; of goby 56; of harvest mouse 132, **133**; of hedgehog 126; of lamprey viii; of mole 126, **127**; of mouse 132, 134; of rabbit 138, **139**; of rat 134; of salmon 104, **105**; of shanny 72; of shrew 128; of squirrel 136, **137** of stickleback 90, **91**; of trout 102, 106, 108; of turtle 120; of vole 130
New Forest 203
New Forest Pony 152, **153**
Newt(s) 114, **115**, 196; larva 114, **115**
Nilsson's Pipe-fish 64, **65**
Noctule Bat 162, **163**
Noemacheilus 194; *barbatulus* 100
North Atlantic Right Whale 190
North Sea **200**
Northern Rockling 76, **77**
Nose-leaf of bat **vii**, 154
Nostril **vii**
Norway Bullhead 66, **67**
Norway Haddock 40, **41**
Norway Pout 36, **37**
Norway Rat 134
Norwegian Topknot 52, **53**
Notochord vi, viii
Nuptial pad **vi**, 116, 118
Nursehound 8
Nutria 130
Nyctalus 197; *leisleri, noctula* 162
Nyctiphanes **21**

Ocean Pipe-fish 64
Oceanic Bonito 26, **27**
Oceanic Perches 40
Ocellus **vii**, 16, **17**
Odd-toed ungulates 197
Odobenidae 180
Odobenus 197; *rosmarus* 180
Odontoceti 182, 197
Odontogadus merlangus 34
Oesophagus vi
Old-wife 28
Omnivore(s) 28, 74, 96, 98, 102, 130–134, 164

Ondatra* 198; *zibethicus* 130
Onos cimbrius, mustela 76
Opah 40, **41**
Operculum **vii**, 20, 193, 194
Ophidia 124, 196
Orcinus 197; *orca* 186
Order 192
Orfe 112, **113**
Orkney Islands **200**
Orkney Vole 130
Oryctolagus 198; *cuniculus* 138
Osmerus 193; *eperlanus* 88
Ostariophysi 92, 96, 98, 112, 193
Osteichthyes 20, 193
Osteoglossomorpha 193
Otolith 44, 58
Otter 174, **175**, 197, 201, 202
Oviparity, in rabbit-fish 18; in ray 12, 14, 16; in reptile 196; in shark 8; in skate 14
Ovipositor **113**
Ovis 197; *aries* 150
Ovoviviparity, in bony fish 40; in lizard 122; in ray 10, 18; in shark 2–6; in snake 124
Oxynotus 192; *centrina, paradoxus* 6

Pagellus 195; *bogaraveo, centrodontus, erythrinus* 28
Pagrus 195; *pagrus* 28
Painted Goby 82, **83**
Painted Ray 16
Palmate Newt 114, **115**
Pandora 28, **29**
Paracanthopterygii 194
Parental care vi, 74, **75**, 78
Parr 104–**107**, 201
Parr marks 104, 106
Parti-coloured Bat 160, **161**
Parturition 162, 184
Paternal care 64, 66, 80, 82, 90, 100
Pearl-side 22, **23**
Pectoral fins, of bony fish **vii**, 40, 62, 68, 193; of cartilaginous fish **vii**, 2, 10, 18, 192, 193
Pedicle(s) **vii**, 146, 148
Pegusa 195; *lascaris* 50
Pelamid 26. **27**
Pelvic fins, of bony fish **vii**, 32, 36, 56, 74, 80, 193, 194; of cartilaginous fish **vii**, 2, 192
Penis 126
Perca 195; *fluviatilis* 94
Perch 94, **95**, 195, 201
Percidae 94

Perciformes 24, 28, 56, 58, 60, 62, 70, 72, 74, 82, 112
Percoidei 28, 30, 42, 58, 94, 195
Perissodactyla 152, 197
Petromyzon 192; *marinus* viii
Pharynx vi
Phoca 197; *groenlandica, hispida* 180; *vitulina* 176
Phocaenidae 182
Phocidae 176–180
Phocoena 197; *phocoena* 182
Pholis 195; *gunnellus* 74
Phoxinus 194; *phoxinus* 100
Phrynorhombus 195; *norvegicus, regius* 52
Phycis 194; *blennoides* 36
Phylum 192
Physeter 197; *catodon* 188
Physeteridae 188
Pig(s) 197
Pigment-containing cells 44, 66, 114, 116
Pigmy Shrew 128, **129**
Pike 110, **111**, 193, 201
Piked Whale 190
Pike-perch 94, **95**
Pilchard 22, **23**
Pilot Fish 30, **31**
Pilot Whale 186, **187**
Pine Marten 170, **171**
Pinnipedia 176–180, 197
Pipe-fish 64, **65**, 194
Piper 68
Pipistrelle Bat 162, **163**
Pipistrellus 197; *nathusii* 158; *pipistrellus* 162
Placenta, of mammal 126, 197; of shark 4, 6
Placodermi 192
Plaice **vii**, 44, **45**, 195
Plain Bonito 24
Plankton 199
Plastron 120
Platichthys 195; *flesus* 46
Plecotus 197; *auritus, austriacus* 160
Pleuronectes 195; *cynoglossus* 46; *platessa* 44
Pleuronectidae 44–48
Pleuronectiformes 44–54, 195
Pogge 80, **81**
Polecat 168, **169**
Pollachius 194; *pollachius, virens* 38
Pollack 38, **39**
Pollan 110, **111**
Polyprion 195; *americanus* 58
Pomatoschistus 195; *americanus* 58;

microps, minutus, pictus 82
Ponies 152
Poor Cod 36, **37**
Pope 94
Population, animal 202, 203
Porbeagle 2, **3**
Porpoise 182, 197
Pouch for young 64
Pouched mammal 197
Poutassou 36
Pouting 36, **37**
Powan 110, **111**
Predator(s) 199, 201, 202, 203;
    protection from 90
Pricket 146
Prionace 192; *glauca* 4
Pristiurus melastomus 6
Protacanthopterygii 193
Prototheria 196
Pseudocalanus 20, **21**
Pseudorca 197; *crassidens* 186
Puffer Fish 42, **43**, 196
Pungitius 194; *pungitius* 90
Pup: seal 176–**179**; shark 4
Pusa hispida 180
Pygosteus pungitius 90

Rabbit(s) 138, **139**, 198, 202, 203
Rabbit Fish 18, **19**, 200
Race 192
Rainbow Trout 102, **103**
Rainbow Wrasse 62, **63**
Raja 193; *alba, batis* 14; *brachyura*
    12; *circularis* 14; *clavata* 12;
    fullonica 16; *marginata* 14;
    microocellata 16; *montagui* 12;
    naevus 16; *oxyrinchus* 14;
    richardsoni, undulata 16
Rajidae 12–16
Ram 150, **151**
Rana 196; *dalmatina, esculenta,*
    ridibunda, temporaria 116
Rangifer 197; *tarandus* 148
Raniceps 194; *raninus* 36
Ranidae 116
Rat(s) 134, **135**, 197, 198, 202, 203
Rat-fish 18, 193
Rattus 198; *norvegicus, rattus* 134
Ray **vii**, 192, 193; electric 10, **11**;
    long-snouted 14–**17**; manta 18;
    short-snouted 12–**17**
Ray's Bream 30
Record rod-caught fish vi
Red Band-fish 30, **31**
Red corpuscles vi

Red Deer **vii**, 142, **143**
Red Gurnard 68, **69**
Red Mullet 30, **31**
Red Sea Bream 28, **29**
Red Squirrel 136, **137**
Redd 106
Redfish 40, **41**
Refection 138
Reindeer 148, **149**
Remora 42, **43**
Remora 195; *remora* 42
Reproduction rate 202
Reptile(s) **vii**, 120–124, 202;
    fossils of 198; ruling 196
Reptilia 120
Respiratory system vi
Reticulated Dragonet 70, **71**
Rhinocerus 197, 201
Rhinolophidae 154
Rhinolophus 197; *ferrumequinum,*
    hipposideros 154
Rhinonemus 194; *cimbrius* 76
Rhodeus 194; *sericeus* 112
Rib-faced Deer 148
Right Whale 190
Ringed Seal 180, **181**
Ringed Snake 124
Risso's Dolphin 184, **185**
River Lamprey **viii**, **1**
Rivers: Rhine, Thames 201
Roach 96, **97**, 201
Rock Cook 62, **63**
Rock Goby 82, **83**
Rocklings 76
Rodentia 130–136, 197
Rodents 130–136, 202
Roe Deer 144, **145**
Roker 12
Romer, A. S. 192
Rorqual 188–190
Rough Hound 8
Rudd 96, **97**, 201
Rudolphi's Rorqual 190
Ruffe 94, **95**
Rut 142–148
Rutilus 194; *rutilus* 96

Sagitta 20, **21**
Saint Kilda 200, 202
Saint Peter's Fish 40
Salamander 196
Salmo 193; *gairdneri* 102; *salar* 104;
    trutta 106, 108
Salmon 104, **105**, 193, 200, 201
Salmonidae 102–108

Salmonids 104–108
Salmoniformes 22, 88, 102–110, 193
Salmonoidei 88, 110, 193
Salvelinus 193; *alpinus* 110;
    fontinalis 102
Sand Eel 20, 44, 60, **61**, 195, 200
Sand Goby 82, **83**, 200
Sand Lizard 122, **123**, 202
Sand Smelt 74, **75**
Sand Sole 50, **51**
Sandy Ray 14, **15**
Sarcopterygii 196
Sarda 195; *sarda* 26
Sardina 193; *pilchardus* 22
Sargasso Sea 84, 86, **87**
Saury Pike 32
Scad 30, **31**
Scald Fish 52, **53**
Scale, of herring vi; of snake **vii**;
    of trout 108
Scaly reptile 196
Scardinius 194; *erythrophthalmus* 96
Schmidt, J. 86
Sciuridae 136
Sciuromorpha 198
Sciurus 198; *carolinensis,*
    vulgaris 136
Scomber 195; *colias, scombrus* 24
Scomberesox 194; *saurus* 32
Scombroidei 24, 26, 195
Scophthalmus 195; *maximus,*
    rhombus 54
Scorpaeniformes 40, 66, 68, 80, 100,
    194
Scotland **200**
Scutes 88, 193
Scyliorhinus 192; *caniculus,*
    stellaris 8
Sea Bream 28
Sea Hen 80
Sea Horse(s) 64, **65**
Sea Lamprey **viii**, **1**
Sea Scorpion 66, 200
Sea Snail 80, **81**, 200
Sea Trout 106, **107**, 200
Seal(s) 176–180, 197, 200, 201
Sebastes 195; *marinus, norvegicus,*
    viviparus 40
Sei Whale 190
Selachii 2–10, 192
Semi-aquatic mammals 128, 130,
    174–180
Seriola 195; *dumerili* 30
Serotine Bat 162, **163**
Serranus 195; *cabrilla* 58
Set, badger's 164

Sewer Rat 134
Sexual dimorphism, of common
    dragonet 70, **71**; of cuckoc wrasse
    62, **63**; of deer 142–148; of goby
    56; of killer whale 186, **187**;
    of lizard 122, **123**; of mail-cheeked
    fish 80, **81**; of minnow 100, **101**;
    of newt 114, **115**; of salmon 104,
    105; of sea horse 64, **65**; of seal
    178; of stickleback 90, **91**;
    of trout 106–**109**
Shad 22
Shagreen Ray 16, **17**
Shanny 72, **73**
Shark(s) **vii**, 2–8, 192, 199
Sharp-nosed Mackerel Shark 2
Sheep 150, 197
Sheltie 152
Shetland Islands **200**
Shetland Pony 152, **153**
Ship Rat 134
Short-finned Tunny 26
Short-snouted rays 12, 14, 16
Short-spined Sea Scorpion 66, **67**
Short-tailed Vole 130, **131**, 202
Shrew(s) 128, 197, 202
Sibbald's Rorqual 190
Sika Deer 146, **147**
Siluriformes 92, 194
Silurus 194; *glanis* 92
Silver Bream 96, **97**
Silver Eel 86, **87**
Siphostoma typhle 64
Six-gilled Shark 2, **3**
Skate(s) 12–14, 193
Skeleton, of bone 20, 193;
    of cartilage viii, 2, 192
Skin vi, 2
Skipper(s) 32, **33**, 194, 199
Sleeper Shark 4
Slow Worm 122, **123**, 202
Small-eyed Ray 16, **17**
Small-headed Sucker 78, **79**
Smallest, deer 144; fin whale 190;
    fish 90; mammal 128
Smelt 88, **89**
Smith, M. 192
Smolt 104, 106
Smooth Hound 6, **7**
Smooth Newt 114, **115**
Smooth Sand Eel 60, **61**
Smooth Snake 124, **125**, 202
Snake(s) **vii**, 120–124, 196, 202
Snake Blenny 72, **73**
Snake Pipe-fish 64, **65**
Soay Sheep 150, **151**

Sole 50, **51**, 195
Solea 195; *solea* 50
Soleidae 50
Solenette 50, **51**
Somniosus 192; *microcephalus* 4
Sorel 146
Sorex 197; *araneus, minutus* 128
Soricidae 128
Sound production, of bats 154–160;
    of bullhead 66; of cod 34;
    of deer 142; of dolphins 184;
    of frog 116; of gurnards 68;
    of hedgehog 126; of newt 114;
    of sea horse 64; of shrew 128;
    of toad 118; of weasel 170
Sow: badger 164
Sowerby's Whale 188, **189**
Spanish Mackerel 24
Sparidae 28
Sparling 88
Sparus 195; *aurata* 28
Spawn, of frog 116, **117**;
    of toad 118, **119**
Spawning behaviour, of amphibia
    114–118; of bitterling 112;
    of blonde ray 12; of bream 96;
    of char 110; of cod 34;
    of dragonet 70; of gudgeon 98;
    of lamprey viii; of plaice 44;
    of salmon 104; of sea scorpion 66;
    of shanny 72; of stickleback 90;
    of trout 108; of tunny 26;
    of wrasse 62
Species vi, 192, 201, 202, 203
Specific name 192
Speckled trout 102, **103**
Spent, coalfish 38, cod 34,
    lamprey viii; ray 10
Sperm Whale 188, **189**
Spermaceti 188
Spermatophore 114
Spermatozoa 114
Sphyrna 192; *zygaena* 8
Spinachia 194; *spinachia* 90
Spine, of bullhead 66; of hedgehog
    126; of ray 12, 18; of shark vii, 6;
    of spined loach 92; of stickleback
    90; of trigger fish 42
Spined Loach 92, **93**
Spiny Anteater 196
Spiny Dogfish vii
Spiracle **vii**, 10, 18, 192, 193
Spondyliosoma 195; *cantharus* 28
Spotted Dragonet 70, **71**
Spotted Goby 82, **83**
Spotted Ray 12, **13**

Spraint 174
Sprat 22, **23**
Sprattus 193; *sprattus* 22
Spur Dogfish 6, **7**
Squalius cephalus 98
Squalus 192; *acanthias* 6
Squamata 122, 124, 196
Squatina 192; *squatina* 10
Squirrel(s) 136, **137**, 197, 198, 202,
    203
Stag: red deer **vii**, 142, **143**;
    sika deer 146, **147**
Stenella 197; *styx* 182
Sterlet 88, **89**
Stickleback(s) 90, 194, 201
Sting Ray 18, **19**
Stizostedion 195; *lucioperca* 94
Stoat 170, **171**
Stomach vi
Stomiatoidei 22, 193
Stone Basse 58
Stone Loach 100, **101**, 201
Straight-nosed Pipe-fish 64, 65
Strait of Dover **200**
Streaked Gurnard 68, **69**
Sturgeon 88, **89**, 193
Suborbital gland **vii**
Subphylum 192
Subspecies 192, 202
Sucker(s) 78, 194; of lamprey viii
Suction organ 80, 194
Summer coat, of blue hare 140, **141**;
    of stoat 170, **171**
Suncus etruscus 128
Sunfish 42, **43**, 196
Sweet William 6, 8
Swim bladder 64, 68
Swimming speed, of barbel 98;
    of bleak 96; of blue whale 190;
    of dolphin 182; of electric ray 10;
    of gurnard 68; of herring 20;
    of salmon 104; of stickleback 90;
    of tunny 26
Synapsida 196
Syngnathidae 64
Syngnathus 194; *acus, rostellatus,*
    typhle 64
Tadpole, of frog 116, **117**; of toad
    118, **119**
Tadpole-fish 36
Tail **vii**, 56; fin **vii**; fluke **vii**, 182, 197
Talpa 197; *europaea* 126
Talpidae 126
Tapir 197
Taurulus 195; *bubalis, lilljeborgi* 66
Teeth, canine **vii**; continuously

growing 130; hinged 38; incisor 28, 130; molar 28; pharyngeal 94; specialized 28; tusk-like 142, 148
Teleostei 193
Temperature vi, 26, 196, 202
Tench 94, **95**, 201
Ten-spined Stickleback 90, **91**
Terrestrial vertebrates 202
Territory, of badger 164; of cat 172; of deer 142, 144; of fish 98, 100; of otter 174; of rabbit 138; of seal 176–178; of vole 130
Tertiary Period 195, 198
Tetraodontiformes 42, 195
Tetraodontoidei 196
Tetrapods 120
Tetrodon lagocephalus 42
Theria 196
Thickback Sole 50
Thick-lipped Grey Mullet 74, **75**
Thin-lipped Grey Mullet 74
Thornback Ray 12, **13**
Thorogobius ephippiatus 82
Three-bearded Spotted Rockling 76, **77**
Three-spined Stickleback 90, **91**
Thresher Shark 4, **5**
Thunnus 195; *alalunga, thynnus* 26
Thymallidae 102
Thymallus 193; *thymallus* 102
Tinca 194; *tinca* 94
Tine **vii**
Toad 118, **119**, 196
Tom cat 172
Tompot Blenny 72, **73**
Toothed whales 182–188, 197, 199
Tope 8, **9**
Topknots 52
Torpedo 10
Torpedo 193; *marmorata, nobiliana* 10
Torsk 36
Trachinoidei 58, 195
Trachinus 195; *draco, vipera* 58
Trachurus 195; *trachurus* 30
Tragus 156
Trigger Fish 42, **43**, 196
Trigla 195; *cuculus, gurnardus, hirundo, lineata, lucerna, lyra, obscura* 68
Triglidae 68
Trigloporus 195; *lastoviza* 68
Trisopterus 194; *esmarkii, luscus, minutus* 36
Triturus 196; *cristatus, helveticus, vulgaris* 114

Tropidonotus natrix 124
Trout 106, 108, 201
True's Beaked Whale 188, **189**
Trygon 193
Tubfish 68
Tunicate 199
Tunny 26, **27**, 195, 199
Turbot 54, **55**, 195
Tursiops 197; *truncatus* 184
Turtle 120, **121**, 196
Tusk, of walrus 180
Twaite Shad 22, **23**
Two-spotted Sucker 78, **79**

Undulate Ray 16, **17**
Ungulate, even-toed 142, 197; odd-toed 152, 197
Urchin 126
Urodela 114, 196
Uterine milk 10
Uterus, of mammal 126, 164, 176; of shark 2, 6

Variegated Sole 50, **51**
Venomous animals: dogfish 6; ray 18; snake 124; toad 118; weever 58
Vertebrae vi, 192
Vertebrate(s) vi, 199–203
Vespertilio 197; *murinus* 160
Vespertilionidae 156–162
Vibrissae vii
Viper 124
Vipera 196; *berus* 124
Viperidae 124
Viviparity, in fish 40, 76, 194; in lizard 122, 123, 202; in reptile 196; in shark 6, 8
Vixen 166
Vole 130, 202
Vulpes 197; *vulpes* 166

Wales **200**
Wall Lizard 122
Walrus 180, **181**, 197
Warren, rabbit's 138
Warty Newt 114, **115**
Water Deer 148
Water Rat 130
Water Shrew 128, **129**, 201
Water Vole 130, **131**, 201, 202
Water-bat 158
Weasel 170, **171**

Weever(s) 58, 195, 200
Wels 92
Welsh Cob 152
Welsh Mountain Pony 152
Whale Shark 4
Whale(s) 182–190, 197, 199; blowing of **189**, **191**; spouting of **189**, **191**
Wheeler, A. 192
Whiskered Bat 156, **157**
White Bream 96
White Cattle 150
White Goby 56
White Skate 14
White Whale 186
White-beaked Dolphin 184, **185**
White-sided Dolphin 184, **185**
White-spotted Smooth Hound 8
White-toothed Shrew 128, **129**
Whitebait 20, 22, 60
Whitefish 110, 201
Whiting 34, 35
Whitling 106
Wild Cat **vii**, 172, **173**, 197
Wild Horse 152
Wild Sheep 150
Wing membrane **vii**, 156, **157**
Wing, of bat 154; of ray 10, 12
Winter coat, of blue hare 140, **141**; of stoat 170, **171**
Witch 46, **47**
Withers **vii**
Wood Mouse 132, **133**
Worm Pipe-fish 64, **65**
Wrasse 62, 195
Wreckfish 58
Wyville Thomson Ridge **200**

Yarrell's Blenny 72, **73**
Yellow Eel 86, **87**
Yellow Gurnard 68, **69**
Yellow-necked Mouse 132, **133**
Yolk stomach 2

Zander 94
Zeiformes 40, 194
Zeugopterus 195; *punctatus* 52
Zeus 194; *faber* 40
Ziphiidae 188
Ziphius 197; *cavirostris* 188
Zoarces 194; *viviparus* 76
Zoarcidae 76